Statistical Inference

An Integrated Bayesian/Likelihood Approach

MONOGRAPHS ON STATISTICS AND APPLIED PROBABILITY

General Editors

F. Bunea, V. Isham, N. Keiding, T. Louis, R. L. Smith, and H. Tong

Monographs on Statistics and Applied Probability 116

Statistical Inference
An Integrated Bayesian/Likelihood Approach

Murray Aitkin
University of Melbourne
Australia

CRC Press
Taylor & Francis Group
Boca Raton London New York

CRC Press is an imprint of the
Taylor & Francis Group, an **informa** business

A CHAPMAN & HALL BOOK

CRC Press
Taylor & Francis Group
6000 Broken Sound Parkway NW, Suite 300
Boca Raton, FL 33487-2742

First issued in paperback 2019

© 2010 by Taylor & Francis Group, LLC
CRC Press is an imprint of Taylor & Francis Group, an Informa business

No claim to original U.S. Government works

ISBN-13: 978-1-4200-9343-8 (hbk)
ISBN-13: 978-0-367-38394-7 (pbk)

Library of Congress Cataloging-in-Publication Data

Aitkin, Murray A.
  Statistical inference : an integrated Bayesian/likelihood approach / Murray Aitkin.
      p. cm. -- (Monographs on statistics and applied probability ; 116)
  Includes bibliographical references and index.
  ISBN 978-1-4200-9343-8 (hardcover : alk. paper)
  1. Inference. 2. Bayesian statistical decision theory. I. Title. II. Series.

  QA279.5.A397 2010
  519.5'42--dc22   2010008250

Visit the Taylor & Francis Web site at
http://www.taylorandfrancis.com

and the CRC Press Web site at
http://www.crcpress.com

I had the good fortune to inherit both mathematical ability and determination (some might say obstinacy) from my father Alex. His exposition of the use of Cuisenaire rods in primary mathematics teaching (Aitkin and Hughes 1970, Aitkin 1975, Aitkin and Green 1984) provided an inspiring example of the value of simplicity in foundational issues, which I have tried to follow in my own professional life. This is sometimes a hard road to follow, especially when innovation is abandoned or derided.

The support of my wife Irit since 1986 has been a wonderful help and benefit, and this book is dedicated to Alex and Irit.

Contents

Preface

What does this book do?

> ... the fact that there are so many contributors and so many different per-
> spectives, means that one has to be courageous or crazy to undertake the
> writing of a book on statistical inference. On the other hand, the subject
> is so important that one would be crazy not to.
> Welsh (1996, p. xv)

In this book I set out an integrated approach to statistical inference using
the likelihood function as the primary measure of *evidence* for statistical model
parameters, and for the statistical models themselves.

To assess the strength of evidence from the data for competing parameter
values or competing models, I use likelihood ratios between the parameter values or the models. To interpret the likelihood ratios I use a Bayesian
approach, which depends in general on only *noninformative* priors that are
widely used in posterior inference about model parameters. (This approach
does not deny the value of subjective or informative priors, only their
necessity.)

Comparison of different statistical models requires a treatment of the unknown model parameters. This problem is usually treated by Bayes factors; I
give a different approach, due originally to Arthur Dempster, which uses the
full *posterior distribution of the likelihood*. This quite small change to standard
Bayesian analysis allows a very general approach to a wide range of apparently different inference problems; a particular advantage of the approach is
that it can use the same noninformative priors.

Another main contribution of the book is to apply the general Bayesian approach to finite population inference developed by Hartley and Rao (1968),
Ericson (1969), and Hoadley (1969), but little used subsequently. This approach, using the multinomial distribution and a noninformative Dirichlet
prior, has the additional benefit of providing a Bayesian version of the "saturated model" against which any parsimonious parametric model can be
compared; the form of model comparison needed is just that developed in
the posterior likelihood approach. The multinomial/Dirichlet approach also
provides a very useful alternative to many "nonparametric" methods.

These two contributions provide a general integrated Bayesian/likelihood
analysis of statistical models.

The book is intended to provide both an alternative to standard Bayesian inference, and the foundation for a course sequence in modern Bayesian theory

at the graduate or advanced undergraduate level. The intended audience for the book is statisticians of all philosophies, but particularly Bayesians. Statisticians and students without some exposure to Bayesian concepts may find some of the ideas difficult, as I assume considerable background in both Bayes and "classical" (frequentist) theory.

The restriction of the book to *evidence* is deliberate: there are already many books on Bayesian and non-Bayesian *decision theory*, and the purpose of this one is less ambitious, but perhaps more relevant scientifically, in providing a detailed prescription for the assessment of statistical evidence.

Chapter 1 gives a brief overview of the competing theories of statistical inference, in terms of the evidence provided by data. Chapter 2 gives the form of Bayes analysis which is used in the main part of the book. Chapter 3 gives Bayesian versions of the one- and two-sample t-tests, and corresponding normal variance tests. Chapter 4 gives a full exposition of the use of the multinomial model and noninformative Dirichlet prior in "model-free" or "nonparametric" Bayesian survey analysis. Chapter 5 extends Chapter 3 to normal regression and analysis of variance. Chapter 6 gives the treatment of binomial and multinomial data, and gives alternative Bayesian analyses to current frequentist "nonparametric" methods. Chapter 7 gives new goodness-of-fit methods for assessing parametric models.

Chapter 8 discusses two-level variance component models and finite mixtures. These models require Markov chain Monte Carlo (MCMC) methods in general (and for finite mixtures in particular). In this book I do not give details of the necessary methods. While many recent Bayesian books integrate R or WinBUGS procedures into the text, I give only examples, other than those in Chapter 8, which can be implemented in any general statistical package which provides random variate generation from the common probability distributions: the emphasis of this book is on the *principles* of Bayesian inference, and especially Bayesian model comparison. This may disappoint some readers, but details of MCMC procedures can be found in many other books. Future editions of this book may include a detailed discussion of MCMC, and other noniterative methods for posterior inference.

In the main text, I use the "we" form to address the reader, to avoid the otherwise overwhelming "I."

Acknowledgements

The book owes much to many statisticians, and to the past support of the Australian Research Council, the UK Social Science Research Council (now the Economic and Social Research Council), and the National Center for Education Statistics and the Institute of Education Sciences of the US Department of Education.

My first exposure to Bayesian inference was in my senior undergraduate year (1960) at Sydney University, when John Hartigan gave me a Bayesian

reading course. I found Fisher's (1956) criticism of the noninvariance of uniform priors decisive, and as the rest of my statistics courses were frequentist,
Bayesian theory did not convince me. I have therefore found the *actual* in-
variance of uniform priors (Geisser 1984) to be a major point in its favor, as
described later in the book.

At the University of Lancaster 1976–79 as a Social Science Research Council professorial fellow, I had the good fortune to be able to travel widely, and
in the process visited Waterloo for George Barnard's retirement celebration.
I was greatly impressed by Jim Kalbfleish's two-volume Springer introduc-
tory book, and especially by the second volume on likelihood and likelihood
inference. Waterloo was well ahead of other universities in its computation
facilities, and the possibility of computing and plotting likelihoods with good
statistical software began my enthusiasm with this approach.

The Social Science Research Council supported (1979–85) a major research
program I directed, at the new Centre for Applied Statistics at Lancaster, on the
analysis of complex social data using various EM algorithm applications. I am
very much indebted to the senior statisticians who visited Lancaster on this
program and contributed greatly to its and to my development, in particular
George Barnard, Jim Berger, Darrell Bock, Steve Fienberg, Nan Laird, Richard
Royall, and Don Rubin. David Sprott also visited Lancaster, on his own grant.

I was also fortunate in the initial appointments I was able to make to the
research staff at Lancaster: Brian Francis, John Hinde, and Dorothy Anderson.
Brian and John have since developed their careers as professors at Lancaster
and Galway. Nick Longford replaced John Hinde and developed multilevel
modeling into an international consulting career. Other outstanding appointments included David Firth and Andy Wood. The Mathematics Department
at Lancaster had talented and enthusiastic statisticians with whom I was able
to establish fruitful collaborations, particularly Joe Whittaker and Granville
Tunnicliffe Wilson.

On foundational issues, I have benefitted at various times from discussions with George Barnard, Peter Green, Bruce Lindsay, Jim Lindsey, Charles
Liu, Sylvia Richardson, Adrian Smith, David Sprott, and Peter Walley. I was
greatly impressed by a remark by Dennis Lindley at an RSS meeting in the
late 1970s: he said that the foundations of statistical inference were so important that everyone should set aside what they were now doing for two years
and together sort out the foundations. Few statisticians have had the time or
the interest to follow Dennis's prescription; however, it has been one of the
benefits of my academic and research life that I have been able to spend even
more time on this subject!

I owe a great deal to Arthur Dempster's work; though he was not able to
visit Lancaster, his unpublished 1974 conference paper (formally published
in 1997), to which I was directed by Don Rubin, has been outstandingly important in the development of the approach in this book. I tried for some
years to combine the "pure" likelihood approach with frequentist theory, but
was persuaded by Richard Royall during his visit to Lancaster that this was
impossible. It was also clear from the published attempts to develop a pure

likelihood theory (Edwards 1972, Royall 1997) that such a theory was limited by the need to treat nuisance parameters in some ad hoc way outside the precepts of the theory. (I made one such ad hoc attempt, with "canonical likelihoods" with John Hinde in Hinde and Aitkin 1986.) Dempster's seminal paper led to my attempt to replace prior integrated likelihoods in the Bayes factor by posterior integrated likelihoods, leading to the posterior Bayes factor (Aitkin 1991).

This attempt at solving the Lindley paradox was not successful and was widely criticized ("using the data twice" became a new Bayesian crime). However, the central role of likelihood in Bayesian, frequentist, and likelihood theory as a measure of strength of evidence, and its posterior distribution first described by Dempster (1974), form the basis of the inferential approach used in the book. This was developed, initially by posterior numerical integration at the University of Western Australia (Aitkin 1997), and generally by posterior simulation in Newcastle UK, with Richard Boys and PhD student Tom Chadwick (Aitkin et al. 2005, Aitkin et al. 2009).

The appearance of the posterior distribution of the likelihood or the deviance in this approach has led to similar criticism. The recent and little-referenced equivalence of the Bayes factor for nested models to the posterior mean of the likelihood ratio between the models (Kou et al. 2005, Nicolae et al. 2008) should silence, or at least mute, this criticism.

The application of the Bayes approach to finite population analysis owes much to my period in Washington DC (2000–2002) as the chief statistician at the Education Statistics Services Institute of the American Institutes for Research (AIR), where I was a senior consultant and adviser to the National Center for Education Statistics (NCES). I am indebted to Jon Cohen, Gary Phillips, Andy Kolstad, and many others at both AIR and NCES, and to Jon Rao and Phil Kott, for detailed discussions of the survey sampling approach and the design-based/model-based controversy.

The controversy over the posterior distribution of the likelihood has made journal publication of the theory and applications difficult. The advancement of this work therefore owes much to editors and referees who have seen its value: I mention particularly David Hand and Wayne Oldford, at different times editors of Statistics and Computing.

In Melbourne, I have benefitted from stimulating interactions with Charles Liu. Although Charles was only briefly supported by the ARC grant which supported this work in Melbourne, he has been a great stimulus to its development, and many of the examples in the book are based on our joint papers, mostly unpublished. The published paper of which he is the lead author (Liu and Aitkin 2008) has a remarkable example of the serious failure of the integrated likelihoods used in the Bayes factor to identify the most appropriate model, among competing models for memory recall. This is given in detail in Chapter 2.

The generality of the posterior likelihood approach, the wide range of its applications, and the continuing difficulties with journal acceptance of these papers have all stimulated and accelerated the writing of this book.

Chapters from early versions of the book have been used in short courses at the University of Lancaster, the Queensland University of Technology, the International Workshop on Statistical Modeling at Cornell University, and the DAG at Dortmund University. I thank Brian Francis, Kerrie Mengersen, Jim Booth, and Goeran Kauermann for their invitations and support, and the participants in the courses for many helpful questions and comments.

Murray Aitkin
Melbourne, November 2009

1

Theories of Statistical Inference

1.1 Example

We begin with a small example. We draw a simple random sample of size 40 from a finite population of 648 families, and for each family record the family income for the previous tax year. From this sample we wish to draw an inference about the *population mean* family income for that tax year. How is this to be done? The sample of incomes, reported to the nearest 1000 dollars, is given in Table 1.1.

Theories of inference can be divided into two classes: those which use the *likelihood function* (defined below) as an important, or the sole, basis for the theory, and those which do not give the likelihood any special status.

Within the first class, there is a division between theories which regard the likelihood as the *sole* function of the data which provides evidence about the model parameters, and those which interpret the likelihood through other factors.

Within the second class, there is a division between theories which take some account of a *statistical model* for the data, and those which are based exclusively on the properties of estimates of the parameters of interest in repeated sampling of the population. Comprehensive discussions of the main theories can be found in Barnett (1999), Welsh (1996), and Lindsey (1996) to which we refer frequently. The discussion here is limited to the *generality* and *simplicity* of the theories. We illustrate these theories with reference to the population income problem.

1.2 Statistical models

Theories which use the likelihood require a statistical model for the population from which the sample is taken, or more generally for the *process* which generates the data. Inspection of the sample income values shows that (in terms of the measurement unit of $1000) they are *integers*, as are the other unsampled values in the population. So the population of size N can be expressed in terms of the *population counts* N_J at the possible distinct integer values of

TABLE 1.1

Family Income Data, in Units of 1000 Dollars

26	35	38	39	42	46	47	47	47	52
53	55	55	56	58	60	60	60	60	60
65	65	67	67	69	70	71	72	75	77
80	81	85	93	96	104	104	107	119	120

income Y_J, or equivalently by the *population proportions* $p_J = N_J/N$ at these values.

A (simplifying) statistical model is an *approximate representation* of the proportions p_J by a *smooth probability distribution* depending on a small number of *model parameters*. The form of the probability function is chosen (in this case of a large number of distinct values of Y) by matching the cumulative distribution function (cdf) of the probability distribution to the empirical cumulative distribution of the observed values. A detailed discussion of this process is given in Aitkin et al. (2005) and Aitkin et al. (2009). We do not give details here, but the matching process leads to the choice of an approximating continuous cdf model $F(y \mid \lambda)$, and corresponding density function $f(y \mid \lambda) = F'(y \mid \lambda)$; the probability p_J of Y_J is approximated by $F(Y_J + \delta/2 \mid \lambda) - F(Y_J - \delta/2 \mid \lambda)$, where δ is the measurement precision (1 in the units of measurement). When the variable Y is inherently discrete on a small number of values, as with count data, the values p_J are approximated directly by a discrete probability distribution model.

The income sample above is clearly *skewed* with a longer right-hand tail of large values, so an approximating model with right skew would be appropriate. The gamma, lognormal, and Weibull distributions are possible choices.

1.3   The likelihood function

Given a simple random sample $\mathbf{y} = (y_1, \ldots, y_n)$ of size n drawn from the population (assumed for the moment to be large compared to the sample), and an approximating statistical model $F(y \mid \lambda)$, the likelihood function $L(\lambda \mid \mathbf{y})$ (of the model parameters λ) is the probability of the observed data as a function of these parameters:[1]

$$L(\lambda \mid \mathbf{y}) = \Pr[y_1, \ldots, y_n \mid \lambda]$$

$$= \prod_{i=1}^{n} [F(y_i + \delta/2 \mid \lambda) - F(y_i - \delta/2 \mid \lambda)]$$

$$\doteq \left[\prod_{i=1}^{n} f(y_i \mid \lambda) \right] \cdot \delta^n$$

if the measurement precision is high relative to the variability in the data.

---

[1] The likelihood is frequently defined to be *any constant multiple* of the probability of the observed data, but in our approach likelihoods *are* probabilities.

In general the parameter vector λ can be partitioned into a subvector θ, of *parameters of interest*, and a subvector ϕ of *nuisance parameters*. We want to draw conclusions about the parameters of interest θ, but the model depends as well on the nuisance parameters ϕ.

For example, we use the gamma distribution with parameters μ (the mean) and r (the *shape* parameter) as the specified model, where μ is the parameter of interest and r is the nuisance parameter:

$$f(y \mid \mu, r) = \frac{r^r}{\Gamma(r)\mu^r} \exp(-ry/\mu)\, y^{r-1}.$$

This parametrization is convenient because it gives orthogonality in μ and r in the information matrix. For the more common parametrization giving the gamma density

$$f^*(y \mid \mu, r) = \frac{1}{\Gamma(r)\mu^r} \exp(-y/\mu)\, y^{r-1},$$

the mean is $r\mu$, which immediately complicates inference about the mean. We discuss this important issue further in Chapter 2.

Then for high measurement precision δ, the gamma likelihood function can be written as

$$
\begin{aligned}
L(\mu, r \mid \mathbf{y}) &= \prod_{i=1}^{n} \left[\frac{r^r}{\Gamma(r)\mu^r} \exp(-ry_i/\mu)\, y_i^{r-1} \cdot \delta \right] \\
&= \frac{r^{nr}}{\Gamma^n(r)\mu^{nr}} \exp(-rT/\mu)\, P^{r-1} \cdot \delta^n
\end{aligned}
$$

where $T = \sum_i y_i$, $P = \prod_i y_i$. An important theoretical point is that the likelihood function here depends on the data through only the two data functions T and P (and the sample size n). These *sufficient statistics* are all that is needed to compute the likelihood function – we do not need the data values themselves.

We will frequently drop for simplicity the $\mathbf{y}$ from the notation for the likelihood function, but it is always implicit in its definition that the data *have been observed*. We now describe briefly the theories and how they deal with inference about the population mean income.

1.4 Theories

1.4.1 Pure likelihood theory

This theory (see for example, Edwards 1972 or Lindsey 1996) uses the likelihood *exclusively* to draw inferences about the model parameters. It is *fully conditional* on the observed data, so denies the relevance of repeated sampling,

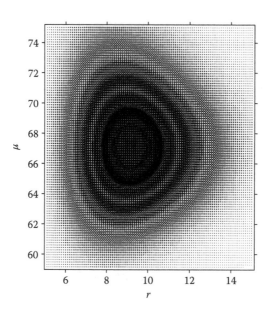

FIGURE 1.1
Gamma likelihood, family income data.

but also explicitly rules out prior distributions for model parameters. Since the likelihood is the probability of the observed data, it can be used as a measure of relative plausibility of different parameter values: the relative evidence for different values λ_1 and λ_2 of the model parameter λ is assessed through their *likelihood ratio* $L(\lambda_1)/L(\lambda_2)$. A likelihood ratio of 10 means that the data support the value λ_1 10 times as strongly as the value λ_2.

So we can compare the data support for pairs of values (μ, r) like $(60, 3)$ and $(65, 3)$, or $(60, 3)$ and $(65, 5)$. Figure 1.1 shows a form of "contour plot" of the two-parameter likelihood in μ and r, by computing the likelihood over a 100-point equally-spaced grid in μ and r, chosen to cover the region of appreciable likelihood. The size of the plotting symbol in the plot is proportional to the likelihood.

In pure likelihood theory, this plot represents the data information about both parameters. For two general values of μ with the same r, (μ_1, r) and (μ_2, r) the likelihood ratio is

$$\frac{L(\mu_1, r)}{L(\mu_2, r)} = \frac{r^{nr} \exp(-rT/\mu_1) \, P^{r-1} \cdot \delta^n/[\Gamma^n(r)\mu_1^{nr}]}{r^{nr} \exp(-rT/\mu_2) \, P^{r-1} \cdot \delta^n/[\Gamma^n(r)\mu_2^{nr}]}$$

$$= \left(\frac{\mu_2}{\mu_1}\right)^{nr} \exp(-rT[1/\mu_1 - 1/\mu_2]).$$

This depends critically on the nuisance parameter r, as the log of the likelihood ratio increases or decreases in magnitude linearly with r. However, the theory *does not specify how nuisance parameters are to be treated*, and so has very restricted application to simple cases where either there are *no* nuisance parameters, or the likelihood function is *separable* in the parameters of interest, that is, it can be expressed as

$$L(\theta, \phi) = L_1(\theta) \cdot L_2(\phi),$$

where L_1 and L_2 are functions, respectively, of θ only and ϕ only. If this is the case, it follows immediately that

$$\frac{L(\theta_1, \phi)}{L(\theta_2, \phi)} = \frac{L_1(\theta_1)}{L_1(\theta_2)},$$

and the nuisance parameter can be ignored: every *section* through the likelihood parallel to the θ axis gives the same likelihood ratio for θ_1 to θ_2, whatever the value of ϕ.

In other cases (the vast majority), some treatment of the nuisance parameters is necessary, and different ad hoc treatments may lead to different conclusions. Lindsey (1996) gave a very detailed discussion of this theory. A common approach is to eliminate the nuisance parameter by *profiling*: replacing the unknown ϕ by $\hat{\phi}(\theta)$, its *maximum likelihood estimate* (MLE) *given* θ. In the case of separable parameters, this replaces ϕ by its MLE $\hat{\phi}$, giving the correct likelihood ratio, but in general $\hat{\phi}(\theta)$ defines a curved path through the parameter space. This process *overstates the information about the nuisance parameter*, and the resulting profile likelihood is *overprecise*.

1.4.2 Bayesian theory

Bayesian theory was dominant (indeed, the *only* theory) from the 1800s to the 1920s. It is *fully conditional* on the observed data, and conclusions about the population from which it was drawn are based on the likelihood function $L(\lambda)$ (representing the data information) and the *prior* (probability) *distribution* $\pi(\lambda)$ of the model parameters, representing the information we have about these parameters external to, and in advance of, the sample data. Inference is expressed through the *posterior* distribution $\pi(\lambda \mid \mathbf{y})$ of the model parameters, *updated* from the prior by the likelihood through Bayes's (also called, as in this book, Bayes) theorem:

$$\pi(\lambda \mid \mathbf{y}) = \frac{L(\lambda)\pi(\lambda)}{\int L(\lambda)\pi(\lambda)d\lambda}.$$

If λ can take just one of the two values λ_1 and λ_2, with prior probabilities π_1 and π_2, the ratio of posterior probabilities (the *posterior odds* for λ_1 to λ_2) is

$$\frac{\pi(\lambda_1 \mid \mathbf{y})}{\pi(\lambda_2 \mid \mathbf{y})} = \frac{L(\lambda_1)\pi_1}{L(\lambda_2)\pi_2}$$
$$= \frac{L(\lambda_1)}{L(\lambda_2)} \cdot \frac{\pi_1}{\pi_2},$$

so that *the posterior odds is equal to the likelihood ratio multiplied by the prior odds.*

So the likelihood ratio plays the same role in Bayesian theory that it does in pure likelihood theory – to provide the data evidence for one parameter value over another – but this is complemented in Bayesian theory by the prior information about these values, their prior probabilities.

The theory requires that we express prior information as a probability distribution. In many cases, we may not have well-developed information or views which are easily expressed as a probability distribution, and much use is made, by many Bayesians, of *weak* or *noninformative* priors, which are "uniformative" relative to the information in the data: the data were presumably collected to obtain information about parameters for which we had little prior information, and so the prior should reflect this lack of information. A noninformative prior for the case of two parameter values would be one with equal prior probabilities, leading to the posterior odds being equal to the likelihood ratio.

A subdivision of Bayes theorists regards noninformative priors as at best undesirable (especially when they are improper), and at worst denying the whole point and advantage of the Bayesian approach, which is to accommodate *both* sample data *and* external information in the same unified probabilistic framework. It argues that *all* prior distributions should reflect the actual information available to the analyst; this may mean that different analysts using different prior distributions come to different conclusions. Analysts who have difficulty formulating priors need to be trained in prior *elicitation* (Garthwaite et al. 2005).

Arguments about the prior, and the meaning, existence, and uniqueness of "noninformative" priors will be discussed more fully in Chapter 2. As stated in Section 1.3, we want in general to draw conclusions about the parameters of interest θ, but the model depends as well on the nuisance parameters ϕ. This is achieved in Bayesian theory by a standard probability procedure: we integrate the *joint posterior distribution* $\pi(\theta, \phi \mid \mathbf{y})$ over ϕ to give the *marginal posterior distribution* $\pi(\theta \mid \mathbf{y})$:

$$\pi(\theta \mid \mathbf{y}) = \int \pi(\theta, \phi \mid \mathbf{y}) d\phi.$$

For the income example, we need to specify the prior distribution for (μ, r). This choice is discussed further in Chapter 2. A simple approach is to specify independent flat priors for μ and r. The posterior density $\pi(\mu, r \mid \mathbf{y})$ is then proportional to the likelihood $L(\mu, r)$.

Figure 1.1, which showed the likelihood, can also be interpreted, for the independent flat priors, as the *joint posterior density* defined over the 100×100 grid of μ_k and r_ℓ for $k, \ell = 1, \ldots, 100$. Strictly speaking it is a posterior *mass function* $\pi(\mu_k, r_\ell \mid \mathbf{y})$ rather than a *density* function. The marginal posterior

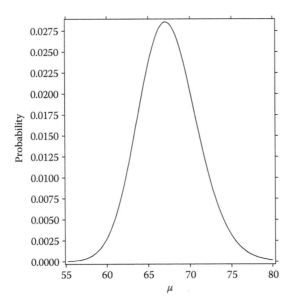

FIGURE 1.2
Posterior mass function, mean income.

mass function of μ is then simply obtained by summing over r_ℓ:

$$\pi(\mu_k \mid \mathbf{y}) = \sum_{\ell=1}^{100} \pi(\mu_k, r_\ell \mid \mathbf{y}).$$

Figure 1.2 shows this mass function, and Figure 1.3 shows the cumulative mass function; these are drawn as continuous curves for visual appeal.

The mean of the posterior mass function $\sum_k \pi(\mu_k \mid \mathbf{y})\mu_k$ is 67.44. Approximate credible intervals and percentiles for the posterior distribution of μ can be obtained by linear interpolation from the cumulative mass function: at $\mu = 60.5$ the cdf is 0.0221 and at 60.75 it is 0.0265, so the approximate 2.5 percentile (to 1 dp) is 60.7; at $\mu = 74.5$ the cdf is 0.9725 and at 74.75 it is 0.9761, so the approximate 97.5 percentile is 74.7; the approximate 95% central credible interval for μ is then [60.7, 74.7].

This simple approach suffers from two difficulties. First, it is limited to small numbers of parameters: with three parameters 100-point grids would require 10^6 points, and the three-dimensional likelihood is difficult to visualize. Second, the marginal distributions of the individual parameters are computed over only 100 points, which does not give sufficient precision for accurate percentiles. We adopt a different approach, by taking advantage of the form of the posterior resulting from the *exponential family* to which the gamma distribution belongs.

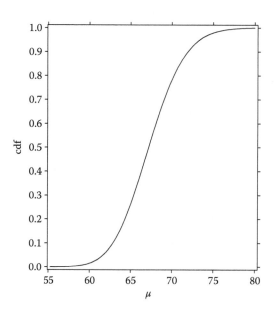

FIGURE 1.3
Posterior cumulative distribution, mean income (1).

We now specify independent flat priors for $\log \mu$ and r; we note below the effect of this change on the posterior distribution for μ. These priors are *improper* (as is the flat prior on μ) – they do not integrate to 1 over the infinite ranges for r and μ – but this does not cause any difficulty if the posterior is proper. The role of r is clear if we reexpress the likelihood in μ and r: writing $\theta = 1/\mu$,

$$L(\mu, r \mid \mathbf{y}) = \frac{r^{nr}}{\Gamma^n(r)\mu^{nr}} \exp(-rT/\mu) \, P^{r-1} \cdot \delta^n$$

$$L(\theta, r \mid \mathbf{y}) = \frac{(rT)^{nr}}{\Gamma(nr)} e^{-rT\theta} \theta^{nr} \cdot \frac{P^{r-1}\Gamma(nr)}{T^{nr}\Gamma^n(r)} \cdot \delta^n.$$

With the prior $\pi(\mu, r) = d\mu dr/\mu$ which is equivalent to $\pi(\theta, r) = d\theta dr/\theta$, the joint posterior distribution of θ and r can be factored into the *conditional* posterior distribution of θ given r, and the *marginal* posterior distribution of r.
The conditional density of θ given r is

$$\pi(\theta \mid r, \mathbf{y}) = \frac{(rT)^{nr}}{\Gamma(nr)} e^{-rT\theta} \theta^{nr-1},$$

a gamma distribution with parameters rT and nr. (It is easily seen that the effect of changing the prior on μ to uniform is to change the prior on θ to $d\theta/\theta^2$; this changes the conditional gamma parameters to $(rT, nr - 1)$). For $n = 40$ and $r > 2$ this makes a negligible difference to the posterior. The

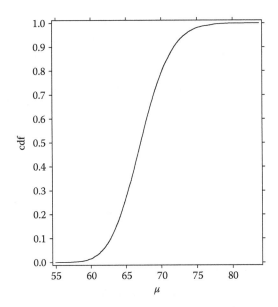

FIGURE 1.4
Posterior distribution, mean family income (2).

marginal density of r is not of any standard form:

$$\pi(r \mid \mathbf{y}) = c \cdot \frac{P^{r-1}\Gamma(nr)}{T^{nr}\Gamma^n(r)},$$

where c is the integrating constant of the density. Analytic integration over r is therefore not possible. However, *simulation* from the marginal density of μ is straightforward and is discussed at length in Chapter 2. We first compute the posterior *mass function* of r over a fine grid, make M random draws $r^{[m]}$ of r from this mass function, and for each m make a random draw $\theta^{[m]}$ from the conditional gamma density of θ given $r^{[m]}$; then random draws $\mu^{[m]}$ of μ are obtained from $\mu^{[m]} = 1/\theta^{[m]}$.

The advantage of this approach is that a much finer grid for r can be used, and the draws of μ are not on a discrete grid; the disadvantage is that the random draws $\mu^{[m]}$, when ordered, define the cdf of μ only up to the random variation inherent in the M simulation values. This is easily assessed: if we wish to estimate percentiles of the posterior distribution of μ, then in M draws of which A draws are less than or equal to μ_A, the estimated cdf at μ_A is A/M, with sampling standard error (in the usual frequentist sense) $\sqrt{A(M-A)/M^3}$. We use this approach extensively in this book.

Figure 1.4 shows the cumulative distribution function of μ for a sample of 10,000 random values of μ generated in this way. It is visually identical to that based on the grid computation.

From the ordered values, we can immediately construct approximations to the *posterior mean of* μ, by simply averaging the 10,000 random values, and to the 95% central *credible interval* for μ, from the 250th and 9750th ordered values. The simulation standard error in the true cdf at these values is $\sqrt{250 \times 9750/10000^3} = 0.0016$, so the standard error of the credibility coefficient (the difference between the true cdfs at the estimated percentiles) is not more than 0.002 (the estimates are positively correlated). This is sufficiently accurate for most purposes: it can be reduced, but not eliminated, by increasing M.

The estimated posterior mean is 67.22, and the approximate 95% central credible interval is [60.5, 74.6]. The credible interval agrees closely with the (interpolated) interval [60.7, 74.7] from the marginal posterior with the flat prior on μ. The posterior means agree less well. We give further details of this approach in Chapter 2.

1.4.3 Likelihood-based repeated sampling theory

This theory was dominant from the 1930s to the 1990s. The theory uses the likelihood function to provide both *maximum likelihood estimates* of parameters and *likelihood-based confidence intervals*, but these are interpreted through their behavior in (hypothetical) *repeated sampling from the same population* (whence the Bayesian term *frequentist* to describe the theory). Prior distributions are not used for single sample inference.

The interpretation of the likelihood function for the income example is now *reversed* compared with the Bayesian interpretation: the parameters μ and r are still fixed constants, but do not have prior distributions. It is the *sufficient statistics* T and P which have the probability distributions, over hypothetical repeated samples drawn from the same population. The same likelihood function

$$L(\mu, r \mid \mathbf{y}) = \frac{r^{nr}}{\Gamma^n(r)\mu^{nr}} \exp(-rT/\mu) \, P^{r-1} \cdot \delta^n$$

can be rewritten differently:

$$L(\mu, r \mid \mathbf{y}) = \frac{r^{nr}}{\Gamma(nr)\mu^{nr}} \exp(-rT/\mu) \, T^{nr-1} \cdot \frac{P^{r-1}\Gamma(nr)}{T^{nr-1}\Gamma^n(r)} \cdot \delta^n,$$

which is very similar to the Bayesian reexpression, but it now refers to the *sampling distributions of* T *and* P. In repeated sampling, T has a gamma($r/\mu, nr$) distribution, and P given T has a distribution determined by the second component.

Without information about r, we cannot use T to draw inferences about μ. The standard approach to this difficulty, in the two-parameter exponential family (see for example, Cox and Hinkley 1974 or Welsh 1996), is to *condition* the repeated sampling inference on the value of the sufficient statistic for the nuisance parameter. Since the statistic on which we are conditioning may have a distribution which depends also on the parameter of interest, it is not clear that we retain full information about the parameter of interest by this approach. For example, if r were the parameter of interest, we would condition on the value of T, to eliminate μ from the likelihood. Since the marginal distribution of T is gamma(r/μ, nr), the conditional likelihood of r given T is simply the second term above:

$$CL(r \mid T) = \frac{P^{r-1}\Gamma(nr)}{T^{nr-1}\Gamma^n(r)}$$

which does not depend on μ, and the log conditional likelihood is

$$\log CL(r \mid T) = (r - 1)\log P + \log \Gamma(nr) - [(nr - 1)\log T + n \log \Gamma(r)].$$

The conditional distribution of $\log P = \sum_i \log y_i$ given T can be obtained, with some effort, from the log-Dirichlet distribution of $\log(y_i/T)$. However, since the distribution of T depends on r, we may have lost information about r in this process.

But we are interested in μ, not r, so we need the conditional distribution of T given P. Since the distribution of P has no simple form, we cannot easily determine the necessary conditional distribution, and r has to be eliminated from the likelihood by other means.

One possible approach is through the profile likelihood (mentioned in Section 1.4.1), in which the unknown r is replaced by its MLE $\hat{r}(\mu)$ given μ, and the resulting profile likelihood in μ is treated as a single-parameter likelihood. Differentiation of the log-likelihood shows that the MLE of r given μ is the solution of

$$\log r - \psi(r) = \mu + \bar{y}/\mu + (\log P)/n - 1,$$

where $\psi(.)$ is the digamma function. Aitkin et al. (2005) and Aitkin et al. (2009) gave detailed discussions of profile likelihoods in a much broader context. The properties of such profile likelihoods are in general not the same as those of single-parameter likelihoods: the replacement of unknown nuisance parameters by functions of the parameter of interest increases the apparent information about the parameter of interest and makes inferential statements based on the profile likelihood overprecise.

Even when the conditional likelihood *can* be determined easily, an obvious question is the *relevance* of the form of conditional sampling. We have to imagine a subfamily of samples in which the sufficient statistic P is exactly the same as in our observed sample, and imagine then the variation in these samples of the other sufficient statistic T. Even in sampling experiments from a *known* gamma population, it would be very difficult to evaluate the properties of the conditional likelihood obtained in this way, since P has a continuous distribution in repeated sampling, and so need never provide the same value in the repeated samples.

The approach to conditioning on the sufficient statistics for nuisance parameters is not universally accepted amongst all frequentists, even when the conditioning statistic is fully *ancillary* – does not depend at all on the parameter of interest. When the distribution of the conditioning statistic depends explicitly on the parameter of interest, the case for conditioning is even less persuasive. For example, in a 2×2 contingency table from two binomial populations, arguments over the appropriateness of the conditional likelihood, obtained by conditioning on the other margin of the table, continue long after Fisher's insistence on the correctness and optimality of the conditioning approach.

We do not need to take a position on this issue, which is often ignored in practice:

> "The choice of the appropriate set of hypothetical repetitions is in principle fundamental, although in practice much less often a focus of immediate concern." Cox (2006, p. 198).

However, there is a close relation between Bayesian and conditional frequentist procedures:

> Pierce (1973) showed that ... good conditional procedures can only be achieved from a Bayesian analysis based on a proper prior, and that good frequentist properties can only hold for procedures which are in a sense limits of proper Bayesian procedures. (Welsh 1996, p. 164)

For the 2×2 table, Altham (1969) showed that Fisher's conditional likelihood arises as a posterior distribution for a specific prior.

The important point for our income example is that the frequentist theory does not provide a straightforward theoretical answer to inference about μ in the gamma distribution. For distributions outside the exponential family (like the Weibull or finite mixtures) the conditioning approach is not available, and other methods of eliminating nuisance parameters have to be found. Grice and Bain (1980) gave approximate small-sample approaches for the gamma mean. In large samples reliance can be placed on the asymptotic normality of functions of the observed data, but this is not a general theory.

1.4.4   "Model-guided" survey sampling theory

The term *model-guided* or *model-assisted* (as used in Särndal et al. 1992) is relatively new in survey sampling theory, which was extensively developed in the 1950s. It refers to the usefulness of model-based *estimators* of model parameters, but without reliance on the *correctness* of the model. Without a formal model, inference is based on the repeated sampling distribution of the *sample selection indicators*, not of the population values themselves. The *survey design* determines the inference, hence the term *design-based* (as opposed to *model-based*) inference.

Our income mean example gives a simple example of the approach. We change notation slightly. Y is the variable of interest, in a finite population of size N. The population values of Y are $Y_1, Y_2, \ldots, Y_N$. The population mean μ is

$$\mu = \sum_{I=1}^{N} Y_I / N$$

and the population variance is

$$\sigma^2 = \sum_{I=1}^{N} (Y_I - \mu)^2 / N.$$

(In the survey literature the variance denominator is usually $N-1$.) We draw a simple random sample without replacement of fixed predetermined size n, and obtain observed values $y_1, \ldots, y_n$, with sample mean $\bar{y}$ and sample variance

$$s^2 = \sum_i (y_i - \bar{y})^2 / (n-1).$$

Define *indicator variables* $Z_1, Z_2, \ldots, Z_I, \ldots, Z_N$: let

$$Z_I = 1 \text{ if population member } I \text{ is selected}$$
$$= 0 \text{ if population member } I \text{ is not selected.}$$

Then

$$\bar{y} = \sum_{i=1}^{n} y_i / n$$
$$= \sum_{I=1}^{N} Z_I Y_I / \sum_{I=1}^{N} Z_I$$
$$= \sum_{I=1}^{N} Z_I Y_I / n.$$

Inference about μ is based on the *repeated sampling properties of the random variable $\bar{y}$ as an estimator of μ*. The fundamental inferential principles are: the Y_I are *fixed constants*, and the Z_I are Bernoulli *random variables*, with

$$\Pr[Z_I = 1] = \frac{\text{no. of samples containing unit } I}{\text{no. of samples of size } n}$$
$$= \frac{\binom{N-1}{n-1}}{\binom{N}{n}} = \frac{n}{N} = \pi,$$

the *sampling fraction*. The properties of $\bar{y}$ are easily established.

$$E[Z_I] = E[Z_I^2] = \pi, \ \text{Var}[Z_I] = \pi(1 - \pi) = \frac{n}{N}(1 - \frac{n}{N}).$$

Hence

$$E[\bar{y}] = \frac{1}{n} \sum_{I=1}^{N} E[Z_I]Y_I$$

$$= \frac{1}{N} \sum_{I=1}^{N} Y_I$$

$$= \mu.$$

So as a random variable, $\bar{y}$ is *unbiased* for μ. For the variance of $\bar{y}$ we need the joint distribution of pairs of the Z_I. These are not independent:

$$Pr[Z_I = 1, Z_J = 1] = Pr[Z_I = 1] Pr[Z_J = 1 \mid Z_I = 1]$$

$$= \frac{n}{N} \cdot \frac{n-1}{N-1}$$

and so

$$Cov[Z_I, Z_J] = \frac{n(n-1)}{N(N-1)} - (\frac{n}{N})^2$$

$$= -\frac{1}{N-1} \frac{n}{N}(1 - \frac{n}{N})$$

$$= -\pi(1 - \pi)/(N - 1)$$

$$\text{Var}[\bar{y}] = \sum_I Y_I^2 \text{Var}[Z_I]/n^2 + \sum_{I \neq J} \sum Y_I Y_J \text{Cov}[Z_I, Z_J]/n^2$$

$$= \frac{1 - n/N}{nN(N-1)}[(N-1)\sum_I Y_I^2 - \sum_{I \neq J}\sum Y_I Y_J]$$

$$= \frac{1 - n/N}{n(N-1)} \sum_I (Y_I - \mu)^2$$

$$= (1 - n/N)\sigma^2/n$$

$$= (1 - \pi)\sigma^2/n$$

if the population variance is defined by the $N - 1$ denominator.

The first term $(1 - n/N) = (1 - \pi)$ is a *finite population correction*: for a small sample fraction, $\text{Var}[\bar{y}] \simeq \sigma^2/n$, but as $\pi \to 1$, $n \to N$, and $\text{Var}[\bar{y}] \to 0$, since the sample exhausts the population.

For the sample variance,

$$E[s^2] = E\left[\sum_i y_i^2 - n\bar{y}^2\right]/(n-1)$$

$$= E\left[\sum_I Z_I Y_I^2 - n\bar{y}^2\right]/(n-1)$$

$$= \left\{n\sum_I Y_I^2/N - n(\text{Var}[\bar{y}] + E[\bar{y}]^2)\right\}/(n-1)$$

$$= \left\{n\sum_I Y_I^2/N - n([1 - 1/N]\sigma^2/n + \mu^2)\right\}/(n-1)$$

$$= (1 - 1/N)\sigma^2.$$

So s^2 is an almost unbiased estimator of σ^2, regardless of any distribution model for Y, and under the Bernoulli model, $\bar{y}$ is the *minimum variance linear unbiased estimator of* μ.

For confidence interval statements about μ, the theory uses the central limit theorem in its general form. The sample mean has expectation μ and variance $(1 - 1/N)\sigma^2$ in repeated sampling, and since it is a (weighted) linear combination of (correlated) random variables Z_I as $n \to \infty$ (and $N \to \infty$), the sampling distribution of

$$z = \frac{\sqrt{n}(\bar{y} - \mu)}{\sigma} \to N(0, 1)$$

as does that of

$$t = \frac{\sqrt{n}(\bar{y} - \mu)}{s},$$

giving the usual large-sample confidence interval

$$\bar{y} - z_{1-\alpha/2}s/\sqrt{n} < \mu < \bar{y} + z_{1-\alpha/2}s/\sqrt{n}.$$

The accuracy of the confidence interval coverage depends on the sample size n – it may be quite inaccurate for small n – and may depend on other properties of the Y population. Without other information about this population, we cannot say more.

For the example, we have $\bar{y} = 67.1$, $s^2 = 500.87$, and the (approximate) 95% confidence interval for the population mean is $[60.1, 74.0]$. Remarkably, we seem to be able to make the same inferential statement about μ without *any* model for the population values Y, or invoking the central limit theorem for the sampling distribution of $\bar{y}$ as a function of $(y_1, \ldots, y_n)$!

However, when we move to regression models, this approach becomes more difficult. The same approach to simple linear regression of Y on X

expresses the usual regression estimator

$$b = \sum_i (y_i - \bar{y})(x_i - \bar{x}) / \sum_i (x_i - \bar{x})^2$$

in the indicator variable form:

$$b = \sum_I Z_I (Y_I - \bar{y})(X_I - \bar{X}) / \sum_I Z_I (X_I - \bar{x})^2$$

with

$$\bar{y} = \sum_{I=1}^{N} Z_I Y_I / \sum_{I=1}^{N} Z_I,$$

$$\bar{x} = \sum_{I=1}^{N} Z_I X_I / \sum_{I=1}^{N} Z_I.$$

Now the repeated sampling distribution of b is much more complex, because it is a ratio of two linear functions of the Z_I, which does not have a simple asymptotic distribution, though by the Mann-Wald theorem the denominator term can be replaced by its expectation in the limiting distribution.

A more important question is *why* we are using the population version of the linear regression in the absence of any model for the population. This is the role of the *guiding* model: if the conditional distribution of Y given X has a linear regression on X with constant variance, the least-squares estimate would be the minimum variance unbiased estimate, and if the conditional distribution of Y given X were in addition normal, the regression estimator would be maximum likelihood and optimal.

So *if* the model were to hold, the sample survey estimator would be identical to the optimal estimator, but if the model does *not* hold, the sample survey estimator should still provide a good estimate of the corresponding population quantity

$$B = \sum_I Z(Y_I - \mu_Y)(X_I - \mu_X) / \sum_I (X_I - \mu_X)^2.$$

From the viewpoint of model-based likelihood theory, this approach is unsatisfactory. The argument is clear if we construct the likelihood as the probability of *all* observed data. The data are *both* the sample selection indicators Z_I *and* the observed response variables y_i for the selected population members, so we need a population model for the Y_I as well.

The fundamental probability relation we use is

$$\Pr[Y_I, Z_I] = \Pr[Z_I \mid Y_I] \Pr[Y_I]$$
$$= \Pr[Y_I \mid Z_I] \Pr[Z_I].$$

where $\Pr[Z_I \mid Y_I]$ is the *sample selection model* for Z – it specifies how the selection probability of population member I depends on the value of the response

Y_I for that member – and $\Pr[Y_I]$ is the *population model* for Y. For simple random sampling,

$$\Pr[Z_I \mid Y_I] = \Pr[Z_I] = \pi^{Z_I}(1-\pi)^{1-Z_I},$$

the Bernoulli model. Correspondingly,

$$\Pr[Y_I \mid Z_I] = \Pr[Y_I]$$

the model for the selected population values is the same as that for the unselected values. So

$$\Pr[Y_I, Z_I] = \Pr[Y_I]\Pr[Z_I],$$

and the likelihood is

$$\begin{aligned}
L &= \Pr[y_1, \ldots, y_n] \cdot \Pr[Z_1, \ldots, Z_N] \\
&= \Pr[y_1, \ldots, y_n] \cdot \frac{1}{\binom{N}{n}} \\
&= \Pr[y_1, \ldots, y_n] \cdot \frac{n}{N}\frac{n-1}{N-1} \cdots \frac{1}{N-n+1}.
\end{aligned}$$

The last term in the selection probabilities is completely known from the design – it is just a constant. In inferential statements about the parameters based on *ratios* of likelihoods, these constant terms *cancel*. Thus, regardless of the kind of model we might have for the Y_I, any inference through likelihood ratios does not depend on the sample design, if this is *noninformative*, in the sense described above – that $\Pr[Z_I \mid Y_I] = \Pr[Z_I]$ – membership of the Ith population member in the sample does not depend on the value of the response Y_I.

In survey sampling theory, this difficulty is countered by the difficulty of the dependence of model-based inference on the correctness of the model – if this is incorrect, the conclusions from the analysis could be wrong. Since every model is by definition wrong (as it is a simplification), the risk of wrong conclusions from the model-based approach is inherent in the approach. Survey samplers are frequently working to a time line for analysis, so they cannot spend much time on investigating and validating suitable models for the response Y. Also, if the sample design is *informative*, likelihood-based inference becomes much more difficult because the form of dependence in $\Pr[Z_I \mid Y_I]$ needs to be specified and included in the likelihood.

In Chapter 4, we address the probability model specification difficulty, using a *minimal* probability model and prior which making no restrictive smooth assumptions – the model is *always* correct!

1.5   Non-model-based repeated sampling

Other data analysis methods have been invented which do not use specific models, or if they do, do not use likelihood-based methods to analyze the data. Some inventors of methods are dissatisfied with the model-based approach – they feel models are *restrictive* and stand in the way of creative, innovative approaches to data analysis. The resulting methods are inherently ad hoc from a theoretical viewpoint because they are not based on general theoretical principles. So their properties have to be assessed by repeated sampling experiments from known data structures, in terms of test sizes or coverage properties of intervals in repeated sampling.

Some methods are closely related to formal statistical models, and when viewed or restructured in terms of models can be improved or evaluated by standard statistical theory. Neural networks are an example: they have received much statistical attention, and the multilayer perceptron or feed-forward neural network has been reexpressed as a formal latent variable model by several statisticians, including Aitkin and Foxall (2003), who pointed out difficulties with the likelihood in the conventional formulation, and gave improved algorithms for fitting the reexpressed model by maximum likelihood.

The danger of repeated sampling inference which ignores the model is clear from the simple case of the uniform distribution with known scale but unknown location. We have a random sample $(y_1, \ldots, y_n)$ from the uniform distribution with density

$$f(y \mid \theta) = 1 \text{ for } y \in (\theta - 0.5, \theta + 0.5)$$
$$= 0 \text{ for } y \notin (\theta - 0.5, \theta + 0.5).$$

A logical analysis can be based on the repeated-sampling distribution of $\bar{y}$, which rapidly approaches $N(\theta, 1/(12n))$. So an approximate 95% confidence interval for θ is $\bar{y} \pm 1.96/\sqrt{12n}$. As an example, here is a sample of 10 from this distribution in which the true θ is 0.9:

```
(0.884, 0.549, 1.126, 1.136, 0.920, 0.770, 1.000, 1.261,
     1.070, 1.047)
```

with mean $\bar{y} = 0.976$, and approximate 95% confidence interval $0.976 \pm 0.179 = [0.797, 1.155]$. The confidence interval will indeed have the correct coverage asymptotically. However the likelihood function is

$$L(\theta) = 1 \text{ for } \theta \in (y_{(n)} - 0.5, y_{(1)} + 0.5)$$
$$= 0 \text{ for other } \theta,$$

where $y_{(1)}$ and $y_{(n)}$ are the smallest and largest order statistics of the sample. So the interval $[y_{(n)} - 0.5, y_{(1)} + 0.5] = [0.761, 1.049]$ is a *100% confidence interval for θ* – we are *certain* that θ lies in this interval, and equally certain that it does not lie outside it! The approximate 95% repeated-sampling interval is longer and contains *impossible* values in the interval $(1.049, 1.155]$, while discounting the well-supported values in the interval $[0.761, 0.796]$.

Further, *there is no preference for one value of θ over another within the 100% confidence interval* – all values of θ are equally well supported. The inference based on the sampling distribution of $\bar{y}$, though formally correct asymptotically in its statement of coverage probability, is irrelevant to the actual information in the data, because it does not recognize the properties of the statistical model for the data. This is not an issue of the failure of regularity conditions (the range of Y does not depend on θ) but a failure (in frequentist terms) to *condition on an ancillary*. A more detailed analysis (Welsh, 1996; pp. 157–9) leads to a similar conclusion for the sampling distribution of *any* estimator of θ, like the mid-range, defined as $(y_{(1)} + y_{(n)})/2$: unless these estimators are *conditioned* on the ancillary statistic (the range $y_{(n)} - y_{(1)}$), similar impossible inferences result.

1.6 Conclusion

Of the theoretical approaches we have considered, only the Bayesian theory is able to deal with the gamma mean inference problem adequately, within its theoretical framework. We now set out the form of the general Bayesian theory we use in the rest of the book.

2

The Integrated Bayes/Likelihood Approach

2.1 Introduction

We use in this book a variation of standard Bayesian theory, for several reasons.

- The theory variation is *general* – it does not need special extensions or – patches to deal with awkward cases. The difficulties of current Bayesian theory, in dealing with point null hypotheses, are resolved by this variation.
- The theory allows us to use noninformative priors *generally* – even with complex models it is not necessary to develop and use informative priors with hyper parameters chosen to fit the data.
- An application of the theory allows us to give a new solution to the vexed question of *model fit* – the adequacy of any model to represent a given set of data.
- Most importantly, the theory is *simple* – a major issue for its understanding and use, and especially for the training of students in statistics.

We develop this version of Bayes analysis in a series of stages. We first discuss some issues of probability which are important for Bayesian analysis.

2.2 Probability

Probability in Bayesian theory has two different functions. In the first, as in frequentist theory, it is used for a *probability model* for data which are variable. So we represent the variation in family income by a gamma distribution, or a lognormal distribution. The most primitive data model is the multinomial, which corresponds to the finite precision of the recorded incomes in the finite population of size N. When we draw a random family from the population and record the family income, the probability that the recorded income is y

is the proportion N_y/N of the value y in the population. (We develop the multinomial model approach fully in Chapter 4.) The likelihood depends on this population probability model.

The second function is to represent our *uncertainty* about model parameters like the population mean. In Bayesian theory a prior distribution is needed for any unknown parameter, and this is updated to the posterior distribution through Bayes theorem and the model likelihood. If the prior distribution is *subjective* or *personal*, that is, is determined by the individual through his or her *beliefs* about this population, based on whatever information he or she may have independently of the data, the posterior distribution will be a *composite* of personal and data-based information.

If the individual's subjective beliefs about the population and its parameters lead to an *informative* prior distribution for the model parameters which is seriously at variance with the information in the likelihood, the posterior will be a composite of inconsistent sources of information. It is for this reason (among others) that the subpopulation of "objective" Bayesians argues for *noninformative* or *weak* priors which provide minimal information about the model parameter; the posterior is then determined essentially by the likelihood, that is by the data. Although the posterior in this case is still formally a measure of personal uncertainty about the model parameter, it is one which most scientists would agree *conveys the information in the data*. Informative priors can then be used in a follow-up analysis; if the conclusions are unchanged, or little changed, from those with the noninformative prior then the data are the main source of information. If the conclusions *are* changed then the prior beliefs are as important as the data, and the conclusions need careful evaluation.

It is important not to interpret the prior as in some sense a *model for nature* – that nature has used a random process to draw a parameter value from a higher population of parameter values, and our job in specifying the prior is to work out what was nature's parameter generating mechanism, and replicate it in the prior. In simple terms, the prior is not a model for a single (parameter) "observation."

The role of the prior is distinct from that of the *upper-level model* in multilevel models for multistage sampling, where we use variance component models to represent the greater homogeneity of the sampled subpopulations at the lower sampling levels. In two-level models, we have an additional level of data requiring a probability model; the unobserved subpopulation parameters at the upper sampling level are given a "random effect" model for their variation which itself depends on parameters like variance components, which play a similar role in multilevel models to the variance in a single-level single-sample problem.

The variance component parameters are given a noninformative prior in our approach. They are not given a full distributional model depending on hyper parameters for the same reason – a single "observation" (parameter) cannot be specified by a probability model: it is a *constant* about which we are uncertain. The prior distribution for the variance components represents our uncertainty about these parameters, and not a higher-level generating process which has produced random realizations of these parameters.

These different roles of probability – as a data model, and as a representation of uncertainty about a parameter – are sometimes called *aleatory* (due to randomness) and *epistemic* (due to uncertainty). O'Hagan (2004) gave a helpful discussion of these concepts.

2.3   Prior ignorance

> All actual sample spaces are discrete, and all observable random variables have discrete distributions. The continuous distribution is a mathematical construction, suitable for mathematical treatment, but not practically observable. E.J.G. Pitman (1979, p. 1).

A long argument over the representation of prior ignorance needs our attention. The use of *uniform* priors for population parameters like proportions and means has been widely criticized since Fisher (1956). The objection is that such priors are not invariant to monotone transformations of the parameter. Suppose our interest is in the *logit* transformation of a population proportion p rather than in p itself, that is in $\theta = \log[p/(1 - p)]$. Using the continuous representation of p, the flat prior is $\pi(p) = 1$ on $(0, 1)$. The corresponding prior distribution of θ is given by

$$\pi(\theta)\mathrm{d}\theta = \pi(p(\theta))\frac{\mathrm{d}p}{\mathrm{d}\theta}\mathrm{d}\theta = e^{\theta}/(1 + e^{\theta})^2 \mathrm{d}\theta$$

which is a logistic distribution, and *not* flat – it has a maximum at $\theta = 0$ and decreases to zero as $\theta \to \pm\infty$. We appear to have a stronger prior belief that θ is near zero – that is that p is near $1/2$ – than that it is large positive or negative.

However, in the actual finite population of size N the population proportion must be one of the numbers $0, 1/N, 2/N, \ldots, N/N$, and a discrete flat prior on p *does* transform to a flat prior on θ:

$$\Pr[p = R/N] = 1/(N + 1)$$

is identical to

$$\Pr[\theta = \log(R/(N - R))] = 1/(N + 1).$$

However, the *values* of θ are no longer equally-spaced – they are closely packed near $\theta = 0$ and sparsely packed for large positive or negative θ. The derivative term $\frac{\mathrm{d}p}{\mathrm{d}\theta}$ in the continuous density accounts for this spacing by giving higher density around $\theta = 0$ and lower density for large values of θ.

So there is no inconsistency in transforming uniform priors to nonuniform scales – the derivative simply defines the measure or "packing density" on the new parameter scale. As Geisser (1984) pointed out,

> Keeping this problem discrete and finite (which in essence it is) avoids the criticism that the uniformity of θ does not imply the uniformity of any function of θ which Fisher (1956, p.16) leveled at Laplace.

We use flat priors (proper or improper) for proportions and means throughout this book, and use commonly accepted "weak" or "diffuse" priors for other parameters, using the diffuse limit of conjugate priors where possible. We do not consider formal rules for deriving noninformative priors (like the Jeffreys priors or the Berger-Bernardo reference priors (Berger and Bernardo, 1989)), because the precise form of prior is unimportant in our approach provided the likelihood dominates the prior in the posterior. This follows from the well-established *principle of precise measurement* or *stable estimation* of Savage (1962, p. 20): for large sample sizes all relatively smooth prior distributions, including diffuse priors, lead to approximately the same posterior distributions.

There are, however, cases in which an apparently reasonable improper prior leads to an improper posterior. A *flat likelihood* is one example: a proper posterior is not *guaranteed* from an improper prior.

2.4   The importance of parametrization

As noted in Chapter 1, we generally have one or more nuisance parameters in the model in addition to the parameter(s) of interest. In marginalizing over the nuisance parameters to obtain the marginal posterior of the parameters of interest, the *association* between the parameters of interest and the nuisance parameters can affect markedly the posterior distribution of the parameter of interest.

If the likelihood is *orthogonal* in the parameters of interest and the nuisance parameters, there is no problem. Since the nuisance parameters are by definition not of interest, we are free to choose the parametrization of the model so that, as far as possible, the parameters of interest are orthogonal to the nuisance parameters. This is generally not possible, and the best we can do is to choose an *information-orthogonal* parametrization in which the (observed or expected) *information matrix* is *block-diagonal* in the parameters of interest and the nuisance parameters. For the gamma distribution considered above, the parametrization in which μ is the mean is already information-orthogonal. This does not give full orthogonality in the parameters; however, as we noted earlier, the likelihood ratio $L(\mu_1, r)/L(\mu_2, r)$ depends strongly on r.

The choice of an orthogonal parametrization was discussed in detail by Cox and Reid (1987). We illustrate its importance with two examples. First a simple example which has been frequently misanalyzed; we follow the treatments in Aitkin and Stasinopoulos (1989) and Aitkin et al. (2005).

2.4.1 Inference about the Binomial N

Given a sample $y_1, \ldots, y_n$ from the binomial distribution $b(N, p)$ with both parameters unknown, what can be said about N? The likelihood is

$$
\begin{aligned}
L_p(N, p) &= \prod_{i=1}^{n} \binom{N}{y_i} p^{y_i} (1-p)^{N-y_i} \\
&= \left[\prod_{i=1}^{n} \binom{N}{y_i} \right] p^T (1-p)^{Nn-T} \\
&= \frac{p^T (1-p)^{Nn-T}}{B(T+1, Nn-T+1)} \cdot \left[\prod_{i=1}^{n} \binom{N}{y_i} \right] B(T+1, Nn-T+1)
\end{aligned}
$$

where $T = \sum_{1}^{n} y_i$, $B(T+1, Nn-T+1)$ is the complete beta function, and the subscript p on L indicates that p is the nuisance parameter.

The structure of the likelihood makes inference about p very simple, as in the gamma distribution for μ. Conditional on N, and with a flat prior on p, p has a posterior beta distribution Beta$(T+1, Nn-T+1)$, while N marginally, with a flat prior, has the nonstandard posterior distribution

$$
\pi(N \mid \mathbf{y}) = c \cdot \left[\prod_{i=1}^{n} \binom{N}{y_i} \right] \cdot B(T+1, Nn-T+1),
$$

which can be computed over an integer grid for N. So we can marginalize over N by making M random draws $N^{[m]}$ from its marginal posterior and making corresponding draws $p^{[m]}$ from Beta$(T+1, N^{[m]}n-T+1)$. However, the term in p in the likelihood contributes substantially to the information about N, so this approach does not work, and we need to eliminate p from the likelihood by marginalizing over p.

This problem was originally considered by Olkin et al. (1981) in the framework of the "instability" of the maximum likelihood estimate (MLE) $\hat{N}$ from samples in the "near-Poisson" region where the sample mean and variance were close.

A recent discussion, with some references, was given by Berger et al. (1999) who argued for the general use of marginal likelihoods for the elimination of nuisance parameters, and gave this model and the following data (considered by Olkin et al. 1981, and Aitkin and Stasinopoulos 1989) as a persuasive example.

The data from a sample of $n = 5$ are 16, 18, 22, 25, 27. The sample mean is $\bar{y} = 21.6$ and the (biased) variance is 17.04, giving moment estimates of $\tilde{p} = 1 - s^2/\bar{y} = 0.211$ and $\tilde{N} = \bar{y}/\tilde{p} = 102.3$. These estimates are highly unstable, as are the MLEs, in the sense that small changes in the largest observation produce very large changes in $\tilde{N}$ – for example, if the largest observation is changed to 28, then $\bar{y} = 21.8$, $s^2 = 19.36$, $\tilde{p} = 0.112$, $\tilde{N} = 194.8$.

2.4.1.1 Profiling

If we eliminate the nuisance parameter p by profiling, the MLE of p given N is $\hat{p}(N) = T/Nn$, which defines a hyperbola $N\hat{p}(N) = \bar{y}$ in the (N, p) parameter space. The profile likelihood in N is then

$$
PL(N) = L_p(N, \hat{p}(N)) = \left[\prod_{i=1}^{n} \binom{N}{y_i} \right] \frac{T^T (Nn - T)^{Nn-T}}{(Nn)^{Nn}}
$$

$$
= \left[\prod_{i=1}^{n} \binom{N}{y_i} \right] T^T (Nn)^{-T} (1 - T/(Nn))^{Nn-T}
$$

$$
\sim \left[\prod_{i=1}^{n} \binom{N}{y_i} \right] \left[\frac{T}{Nn} \right]^T e^{-T}.
$$

On expanding the binomial coefficients using Stirling's formula (with N large compared to n and the y_i), we have

$$
\prod_{i=1}^{n} \binom{N}{y_i} \sim \frac{N^T}{(2\pi)^{n/2} e^{-T} \prod_i y_i^{y_i+1/2}},
$$

and hence for large N,

$$
PL(N) \sim \frac{T^T}{(2\pi)^{n/2} n^T \prod_i y_i^{y_i+1/2}},
$$

which does not depend on N at all – the profile likelihood in N (Figure 2.1) is flat for large N, with a very poorly defined maximum at $N = 100$.

2.4.1.2 Conditioning

If we eliminate p by conditioning on T, the sufficient statistic for p given N, we have

$$
CL(N) = \left[\prod_{i=1}^{n} \binom{N}{y_i} \right] p^T (1-p)^{Nn-T} / \left[\binom{Nn}{T} p^T (1-p)^{Nn-T} \right]
$$

$$
= \left[\prod_{i=1}^{n} \binom{N}{y_i} \right] / \binom{Nn}{T}
$$

$$
\sim \frac{T^{T+1/2}}{(2\pi)^{(n-1)/2} n^T \prod_i y_i^{y_i+1/2}}
$$

and the conditional likelihood conditioned on T has no internal maximum at all, approaching its maximum as $N \rightarrow \infty$ (Figure 2.2).

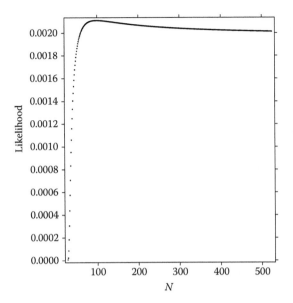

FIGURE 2.1
Profile likelihood.

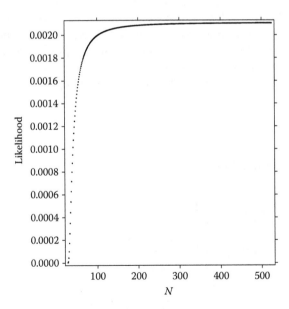

FIGURE 2.2
Conditional likelihood.

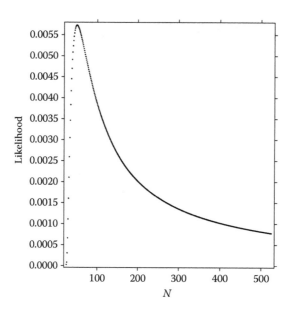

FIGURE 2.3
Integrated likelihood.

Berger et al. (1999) concluded that "These [likelihoods] are nearly constant over a huge range of N and are clearly useless for inference." They proposed the uniform or Jeffreys priors for this problem; these give well-defined modes in the integrated likelihood for N. Figure 2.3 shows the integrated likelihood for the uniform prior; the mode is at $N = 51$.

Kahn (1987) had earlier considered general conjugate beta priors

$$\pi(p) = \frac{p^{a-1}(1-p)^{b-1}}{B(a, b)}$$

and showed that the integrated likelihood in N,

$$IL_{a,b}(N) = \left[\prod_{i=1}^{n} \binom{N}{y_i} \right] \cdot B(T + a, Nn - T + b)$$

$$\sim \left[\prod_{i=1}^{n} \binom{N}{y_i} \right] \cdot \frac{\sqrt{2\pi}(T + a)^{T+a-1/2}}{(Nn)^{T+a} e^{T+a}}$$

$$\sim \frac{(T + a)^{T+a-1/2}}{(2\pi)^{(n-1)/2} n^{T+a} e^a \prod_i y_i^{y_i+1/2}} \cdot N^{-a}.$$

This does not depend at all on the second beta index b, but depends critically (in N) on the first index a, which controls the location of the mode and the

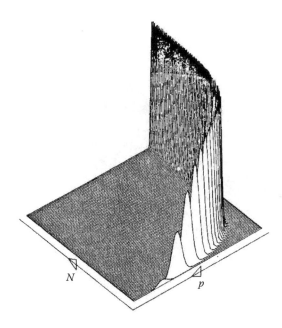

FIGURE 2.4
Likelihood in N and p.

heaviness of the tail of the posterior distribution of N; for $a = 0$ this tail is flat, giving an essentially uninformative posterior for N, equivalent to the conditional likelihood. For $a = 1/2$ (as in the Jeffreys prior) or 1 (as in the uniform prior) the integrated likelihood decreases with large N and has a well-defined interior mode. So the choice of the a parameter of the beta prior for p has a critical effect on the posterior in N.

A detailed comparison of profile likelihood and integrated likelihood inference for this example was given in Aitkin and Stasinopoulos (1989), who also showed the likelihood in N and p, which has extremely concentrated banana-shaped contours along the curve $Np = T/n$ (Figure 2.4).

This very strong association between the parameters emphasizes the difficulty of drawing marginal inferences about N, at least in this parametrization. Aitkin and Stasinopoulos (1989) derived the (observed and expected) information-orthogonal nuisance parameter transformation $\psi = Np$, the mean number of successes, by solving a partial differential equation, following Cox and Reid (1987). In this parametrization we have

$$L_\psi(N, \psi) = \left[\prod_{i=1}^{n} \binom{N}{y_i} \right] \cdot \left(\frac{\psi}{N} \right)^T \left(1 - \frac{\psi}{N} \right)^{Nn-T}.$$

The joint likelihood in N and ψ, shown in Figure 2.5, is almost orthogonal.

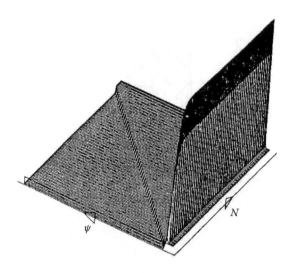

FIGURE 2.5
Likelihood in N and ψ.

For any section across the likelihood at a given ψ, the likelihood ratio for two values of N is

$$\frac{L_\psi(N_1, \psi)}{L_\psi(N_2, \psi)} = \frac{\left[\prod_{i=1}^{n}\binom{N_1}{y_i}\right] \cdot \left(\frac{\psi}{N_1}\right)^T \left(1 - \frac{\psi}{N_1}\right)^{N_1 n - T}}{\left[\prod_{i=1}^{n}\binom{N_2}{y_i}\right] \cdot \left(\frac{\psi}{N_2}\right)^T \left(1 - \frac{\psi}{N_2}\right)^{N_2 n - T}}$$

$$\sim \frac{\left[\prod_{i=1}^{n}\binom{N_1}{y_i}\right]}{\left[\prod_{i=1}^{n}\binom{N_2}{y_i}\right]} \cdot \left(\frac{N_2}{N_1}\right)^T \cdot \frac{e^{-n\psi}}{e^{-n\psi}},$$

which does not depend on ψ. Thus for large N, the parameters N and ψ are separable, and so the likelihood in N behaves like

$$L_\psi(N, \hat{\psi}) = L_\psi(N, T/n) = L_p(N, \hat{p}(N)),$$

the profile likelihood in N, which is invariant over reparametrization.

This is in accord with the "near-Poisson" nature of the sample – the maximized likelihood ratio for Poisson to "best binomial" is 0.935 (Aitkin and Stasinopoulos 1989) – and with the profile likelihood which exhibits this asymptotic behavior.

We note finally that independent flat priors on p and N transform to a prior in N and ψ of the form $\pi(N, \psi) = 1/N$, and it is this term in $1/N$ which "pulls down" the flat N tail in the N, ψ parametrization, giving the appearance of a well-defined mode in N in the N, p parametrization with the uniform prior in p.

The "useless" profile or conditional likelihoods are in fact conveying correctly the information *in the data* about N – the well-defined modes in the integrated likelihoods for the uniform and Jeffreys priors are direct consequences of these priors, and give a misleading impression of the information *in the data* about N.

Thus the use of independent flat priors may have a strong effect on the marginal posterior if the parameters are strongly associated in the likelihood.

2.4.2 The effect size

A second, less obvious, example is the *coefficient of variation* σ/μ for a normal variable. This also appears, as the reciprocal $\theta = \mu/\sigma$, in studies of the *effect size* (more commonly in the two-sample problem, as $(\mu_1 - \mu_2)/\sigma$). For the single-sample normal model, the parametrization in terms of μ and σ is already information-orthogonal. There is no exact distribution for θ, but we have a simple simulation distribution: with the usual noninformative prior $1/\sigma$, we have

$$\mu \mid \sigma, \mathbf{y} \sim N(\bar{y}, \sigma^2/n)$$
$$RSS/\sigma^2 \mid \mathbf{y} \sim \chi_{n-1}^2$$
$$\theta = \mu/\sigma \mid \sigma, \mathbf{y} \sim N(\bar{y}/\sigma, 1/n).$$

The simulation is

- Make M draws $\sigma^{2[m]}$ from the RSS/χ_{n-1}^2 posterior distribution of σ^2;
- For each $\sigma^{[m]}$, make one draw $\theta^{[m]}$ from $N(\bar{y}/\sigma^{[m]}, 1/n)$.

The credible interval for θ or $1/\theta$ is then easily constructed. As an example, suppose that a sample of $n = 25$ gives a sample mean of $\bar{y} = 10$ with $RSS = 96$, $s^2 = RSS/24 = 4$, $s = 2$. Figure 2.6 gives the posterior distribution for θ with the above prior from $M = 10{,}000$ draws.

The median and 95% central credible interval for θ are 4.94, [3.56, 6.46]. The 10,000 draws for θ are graphed against those for σ in Figure 2.7.

A strong negative correlation is visible between θ and σ, not surprising since θ has σ in the denominator! Should this cause us concern? An information-orthogonal parametization exists for the nuisance parameter in this model (Tibshirani, 1989). Figure 2.8 graphs the same θ draws against the orthogonal parameter $\phi = \mu^2 + 2\sigma^2$.

The graph shows very low correlation, so marginalizing to θ in this parametrization gives nearly complete independence of θ and ϕ. But since the same marginal posterior for θ results from both the graphs, the conclusions from the strongly correlated joint posterior are not affected by the correlation.

Further evidence comes from a comparison of the *profile likelihood* for θ with the above posterior; the profile likelihood is straightforwardly calculated. Figure 2.9 shows the profile *relative* likelihood, that is, the profile likelihood normalized to have maximum 1.

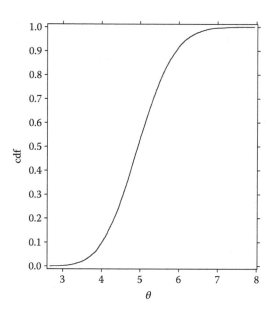

FIGURE 2.6
Posterior: effect size.

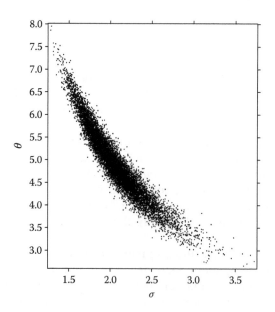

FIGURE 2.7
Joint posterior: effect size and sigma.

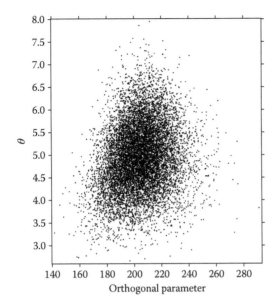

FIGURE 2.8
Joint posterior: effect size and orthogonal parameter.

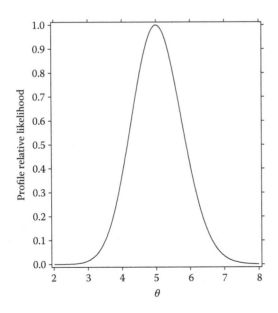

FIGURE 2.9
Profile relative likelihood: effect size.

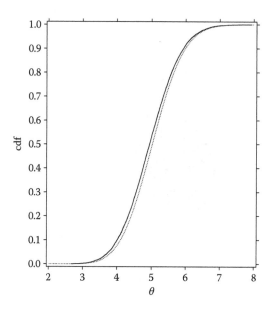

FIGURE 2.10
Profile relative likelihood and posterior: effect size.

Figure 2.10 shows the cdfs for both the profile and posterior; the profile likelihood has been computed as a normalized discrete distribution on a 300-point grid and cumulated.

These distributions agree closely; since the profile likelihood is invariant to the nuisance parametrization this reassures us that the posterior for θ is conveying the information in the likelihood correctly.

2.5 The simple/simple hypothesis testing problem

We now consider the fundamental issue of the comparison of models. This is a controversial area:

> We also considered how far inferences and decision could be based on the numerical values of likelihood ratios. But while we first obtained the critical or rejection regions of our theory as the contours in the sample space on which the appropriate likelihood ratio was constant, we thought that the meaning of the test was more easily grasped if expressed in terms of the probability density inside (or beyond) the boundary of the region, rather than in terms of the likelihood ratio on the boundary.
> Egon Pearson (1962)

* * *

Suppose that a statistician has to make n tests of this kind, each of a simple hypothesis H_0 against a simple alternative H_1, the experiment and the hypotheses in one test having no connection with those in another. Suppose also that for the average size of his tests . . . he wishes to maximize the average power. . . . It can be shown (Pitman, 1965) that to do this, he must use the same critical value c for the likelihood ratio in all the tests. Pitman (1979, p. 3)

* * *

. . . The chapter [1] ends with the dictum, 'There is no room for adhockery in Bayesian statistics' (p. 19). Yet many of the Bayesian estimators proposed later in the book, including Bayes linear estimators (p. 158) and solutions for Bayes factors from improper prior distributions (pp. 183–196) are equally ad hoc, violate the likelihood principle, have been proven to inappropriately use the data twice, are inconsistent, and do not agree with each other. It is unclear whether the latter isolated foray of research represents a wrong turn in Bayesian statistics, and also whether the reader just learning Bayes should be exposed to it. (Ankerst 2005)

We begin with a simpler version of the model comparison problem in Chapter 1. We have two possible models for the data $\mathbf{y}$:

- Model 1: y has a normal distribution $N(\mu_1, \sigma^2)$.
- Model 2: y has a normal distribution $N(\mu_2, \sigma^2)$.

where $\mu_1 = 0$, $\mu_2 = 1$, and $\sigma = 1$. A simple random sample of $n = 25$ gives a sample mean $\bar{y} = 0.4$. Given the data $\mathbf{y}$, how do we decide which model is better supported?

Both Bayesian and frequentist theories give the same answer: compute the likelihood ratio $LR_{12} = L(\mu_1)/L(\mu_2)$; if LR_{12} is "large," the data support Model 1 over Model 2; if it is "small," the data support Model 2 over Model 1. The theories differ in how they decide on "large" and "small."

2.5.1 Bayes calibration

In Bayesian theory, with equal prior probabilities for the two models, the posterior odds are equal to the likelihood ratio, so if this is greater than 1, Model 1 is better supported while if it is less than 1, Model 2 is better supported. The posterior probability of Model 1 is then (with equal prior probabilities) $\pi_{1|y} = LR_{12}/(1 + LR_{12})$. Likelihood ratios LR_{12} of 1, 3, 9, 19, and 99 with equal prior probabilities give posterior probabilities of Model 1 of 0.5, 0.75, 0.9, 0.95, and 0.99.

We will generally take a likelihood ratio of 9 or 10 as reasonably persuasive sample evidence in favor of the numerator model, though this may be outweighed in the posterior probability by strong prior evidence against the model. Since our approach is based on the premise that the data should be able to "tell the story" with weak or noninformative priors, we will generally assume that models being compared are equally well-supported a priori.

For our example the likelihood ratio is

$$LR_{12} = \exp\left\{-\frac{n}{2\sigma^2}[(\bar{y} - \mu_1)^2 - (\bar{y} - \mu_2)^2)]\right\}$$
$$= \exp\left\{-\frac{n}{\sigma^2}(\mu_2 - \mu_1)[\bar{y} - (\mu_1 + \mu_2)/2]\right\},$$

so that the data support Model 1 better if $\bar{y} < (\mu_1 + \mu_2)/2$ – that is, if the sample mean is closer to μ_1 than to μ_2, and Model 2 is better supported if $\bar{y} > (\mu_1 + \mu_2)/2$ – the sample mean is closer to μ_2 than to μ_1.

2.5.2 Frequentist calibration (fixed sample size)

However, the frequentist test with a *fixed sample size* is different. We have to specify one hypothesis as the *null*, and the other as the *alternative*; we then set the *critical region* of the test to ensure a fixed test size (rejection probability of the true null hypothesis). If Model 1 is the null hypothesis, and the (one-sided) test size is set at 5%, this hypothesis is rejected if $\bar{y} > \mu_1 + 1.64\sigma/\sqrt{n}$, which for $n = 25$ means $\bar{y} > 0.328$. So for our sample mean of $\bar{y} = 0.4$, the null hypothesis of Model 1 is rejected.

Under the alternative hypothesis, the probability of rejection is

$$\Pr[\bar{y} > \mu_1 + 1.64\sigma/\sqrt{n} \mid \mu_2] = \Pr[\sqrt{n}(\bar{y} - \mu_2)/\sigma > \sqrt{n}\Delta + 1.64]$$
$$= 1 - \Phi(\sqrt{n}\Delta + 1.64)$$
$$= 1 - \Phi(-3.36)$$
$$= 0.9996$$

where $\Delta = (\mu_1 - \mu_2)/\sigma = -1$. The power of the test is satisfyingly high: if the null hypothesis of Model 1 is false, we will almost certainly reject it in favor of the alternative Model 2.

However, if the null hypothesis is taken to be Model 2, the test procedure will be reversed if we set the (one-sided) test size again at 5%: now Model 2 will be rejected if $\bar{y} < \mu_2 - 1.64\sigma/\sqrt{n} = 0.672$, and the power will be 0.9996 against the alternative Model 1. Now we reject the null hypothesis of Model 2!

So our conclusions depend on which model is declared to be the null, and which the alternative. In either case the *total error probability*, the sum of the test size (the Type I error probability) and the Type II error probability (1 − the power) is 0.0504.

These results are discomforting from two different viewpoints. First, when Model 1 is declared to be the null hypothesis, the likelihood ratio at $\bar{y} = 0.4$ is 12.18 in favor of Model 1, giving, with equal prior probabilities, a posterior probability of Model 1 of 12.18/13.18 = 0.924. We are rejecting the model which is very much better supported.

Second, if we use a likelihood ratio of 1 as the critical value, and reject Model 1 if $LR_{12} < 1$, the size and the Type II error probability of the test are equal, since

$$
\begin{aligned}
\Pr[LR_{12} < 1 \mid \mu_1] &= \Pr[\bar{y} > (\mu_1 + \mu_2)/2 \mid \mu_1] \\
&= \Pr[\sqrt{n}(\bar{y} - \mu_1)/\sigma > -\sqrt{n}\Delta/2] \\
&= 1 - \Phi(-\sqrt{n}\Delta/2) \\
&= 0.0062, \\
\Pr[LR_{12} > 1 \mid \mu_2] &= \Pr[\bar{y} < (\mu_1 + \mu_2)/2 \mid \mu_2] \\
&= \Pr[\sqrt{n}(\bar{y} - \mu_2)/\sigma < \sqrt{n}\Delta/2] \\
&= \Phi(\sqrt{n}\Delta/2) \\
&= 0.0062.
\end{aligned}
$$

So the total error probability with this critical value of the likelihood ratio is 0.0124, far below the total error probability of 0.0504 using the fixed test size of 5%. By setting a test size of 0.0062, we guarantee a power of 0.9938 against the alternative, *whichever way we specify the null and alternative hypotheses*. Why set a much higher Type I error probability, with almost no improvement in power?

The conclusion seems unavoidable: *when testing a simple null against a simple alternative hypothesis, setting an arbitrary fixed test size α without reference to the alternative may lead us to reject the wrong hypothesis.* To prevent this, and to minimize the total error probability, we need to use the likelihood ratio as the test criterion, and (in this symmetric case) equate the two error probabilities.

These results are not peculiar to the normal models in the example – they hold for the comparison of *all* completely specified models. We use the normal models for illustration because the size and power computations are simple for these models.

2.5.3 Nonfixed error probabilities

A further consequence of this use of a likelihood ratio of 1 as the model choice criterion is that *the Type I and Type II error probabilities P are a decreasing function of sample size n and of Δ.* As n and/or $\Delta \to \infty$, both error probabilities tend to zero. We can make model choice with *fully specified models* as precise as we want by taking a sample large enough for the "distance" between the models.

Correspondingly, *in small samples the error probabilities may rise far above conventional levels.* The normal example provides a simple illustration. For the critical likelihood ratio of 1, both error probabilities are given by $\Phi(\sqrt{n}\Delta/2)$. These are shown in Table 2.1 for two values of Δ and seven values of n.

For well-separated alternatives with "effect size" $\Delta = 1$, the error probabilities go rapidly to zero. For poorly separated alternatives with effect size $\Delta = 0.1$, very large sample sizes are necessary to give a clear decision between the alternatives.

TABLE 2.1

Error Probabilities P

Δ	0.1						
n	1	4	9	16	25	36	100
P	0.4801	0.4602	0.4404	0.4207	0.4013	0.3281	0.3085
Δ	1						
n	1	4	9	16	25	36	100
n	0.3085	0.1587	0.0668	0.0228	0.0062	0.0013	0.0000

2.5.4 Frequentist test (sequential sampling)

The frequentist sequential sampling test (Wald 1947) is again different. Here the sample size is not fixed, but is increased sequentially until a decision is made. For a simple $H_1 : \theta = \theta_1$ against a simple $H_2 : \theta = \theta_2$, for a model $f(y \mid \theta)$ with likelihood $L(\theta)$, the user specifies a size α and power $1 - \beta$ of the test. Then at sample size n, the test rejects H_1 and accepts H_2 if the likelihood ratio

$$L R_{12} = \frac{L(\theta_1)}{L(\theta_2)} < \frac{\alpha}{1 - \beta},$$

and rejects H_2 and accepts H_1 if

$$L R_{12} > \frac{1 - \alpha}{\beta}.$$

If

$$\frac{\alpha}{1 - \beta} < L R_{12} < \frac{1 - \alpha}{\beta},$$

the sample is increased to $n + 1$, and this process continues until one hypothesis or the other is accepted. We note for later reference that if $\beta = \alpha$, the "critical values" for decision are symmetric: $\alpha/(1 - \alpha)$ and $(1 - \alpha)/\alpha$. So if we were equally concerned about Type I and Type II errors, we would use a symmetrical decision criterion.

This test can save substantially (in sample size) on the fixed sample size test. Remarkably, it does not need the sampling distribution of the likelihood ratio under either model to set critical values: these are simply determined from the specified size and power.

Suppose we specify the normal example above, with $\mu_1 = 0$, $\mu_2 = 1$, $\sigma = 1$. Assume we want size $\alpha = 0.05$ and power $1 - \beta = 0.8$. Then we proceed to draw observations, evaluating the likelihood ratio at each draw. We express the likelihood ratio in the form

$$L R_{12} = \exp \left\{ -\frac{n}{2\sigma^2} [(\bar{y} - \mu_1)^2 - (\bar{y} - \mu_2)^2)] \right\}$$
$$= \exp[-(z_1^2 - z_2^2)/2]$$

where $z_j = \sqrt{n}(\bar{y} - \mu_j)/\sigma$ is the frequentist "z-statistic" for testing consistency of the sample mean $\bar{y}$ with the hypothetical mean μ_j. We continue sampling as long as

$$0.0625 = 0.05/0.8 < \exp[-(z_1^2 - z_2^2)/2] < 0.95/0.2 = 4.75.$$

This is equivalent to continuing while

$$-3.116 < -2 \log L R_{12} = z_1^2 - z_2^2 < 5.545.$$

At $n = 25$, we have $\bar{y} = 0.4$, so that $z_1 = 2$, $z_2 = 3$, $z_1^2 - z_2^2 = -5$. In terms of the specified α and β, this is very convincing evidence in favor of the null hypothesis that $\mu = 0$.

The conclusions may depend on the setting of α and β (as they do for α in the fixed sample size case), but *these* depend on the specification of *which hypothesis is the null*. As with the fixed sample size test, this specification may lead to difficulties if the models being compared have equal status.

We can avoid this, as in the fixed sample size case, by having *equal Type I and Type II error probabilities*. In the example above, we are accepting a total error probability of 0.25. If we equate the error probabilities for the same total, we could test symmetrically with $\alpha = \beta = 0.125$. This would give critical likelihood ratios of $1/7$ and 7, or critical values of $-2 \log L R_{12}$ of ± 3.89. With a likelihood ratio of 12.18, the data as before clearly point to accepting the hypothesis that $\mu = 0$.

2.5.5 Bayesian interpretation of type I error probabilities

From a Bayesian point of view, with equal prior probabilities on the hypotheses, the equal-error-probabilities approach would correspond in this example to calibrating the posterior probability of each hypothesis as convincing ("accept" this hypothesis) if it is greater than $7/8 = 0.875 = 1 - \alpha = 1 - \beta$. If the probability of neither hypothesis reaches this level, we do not regard the data as providing convincing evidence to discriminate the models. This differs from the usual Bayesian approach only in degree: the common Bayesian calibration of likelihood ratios (or Bayes factors, discussed in Section 2.8) is that a LR of 3 or $1/3$ is barely worth considering as evidence, one of 9 or $1/9$ is suggestive evidence, and one of 30 or $1/30$ is strong evidence. These calibrations correspond to sequential values for $\alpha = \beta$ of 0.25, 0.1, and 0.032.

However, the Bayesian calibration does not need to use fixed values of the likelihood ratio to define "acceptance" or "rejection." *Decisions* or *actions* determined by evidence require costs or losses associated with these actions, which are beyond the scope of our *evidence* treatment.

The use of a 5% test size without reference to the alternative in the fixed sample size case could be interpreted in a Bayesian framework as a case in which the prior probabilities of the two models are *not* equal. Since the data support the "null" hypothesis (Model 1) much more strongly than the "alternative" Model 2 (a likelihood ratio of 12.18), this must mean that we have very strong prior evidence or opinion about the models which outweighs the data evidence. If we set $\pi_1/\pi_2 < 1/12.18 = 0.082$, which means that $\pi_1 < 0.076$, the posterior probability of Model 1 will be less than 0.5, and Model 2 will be (just) better supported by the data, giving conclusions consistent to some extent with the 5% test size.

So the frequentist fixed-sample-size conclusion with the 5% test size corresponds to a Bayes analysis with a *very* strong prior weight against Model 1. This contradicts the idea that the null hypothesis is to be *protected* – that it should be rejected only when the data are inconsistent with it. The heavy prior weight *against* the null hypothesis is not protecting it – it is *exposing* it to rejection from weak, and in this example contradictory data evidence.

We argue that, for model comparisons with fully specified models, we should be using the likelihood ratio as the data criterion for model choice, and not using a fixed test size or Type I error probability to set the critical value for the test.

However, comparisons of fully specified models are rare – we almost always have *nuisance parameters* in one or both models. We generalize the approach above to deal with this in the next section.

2.6   The simple/composite hypothesis testing problem

We modify the previous model comparison problem to the more common case when the alternative is composite. We have two possible models for the data **y**:

- Model 1: y has a normal distribution $N(\mu_1, \sigma^2)$.
- Model 2: y has a normal distribution $N(\mu, \sigma^2)$.

where $\mu_1 = 0$ and $\sigma = 1$, but μ is unspecified, except that it is not μ_1. Given the data y with $n = 25$ and $\bar{y} = 0.4$, how do we decide which model is better supported?

2.6.1   Fixed-sample frequentist analysis

The fixed-sample frequentist theory uses the *ratio of maximized likelihoods* to derive a test statistic, whose sampling distribution under the null hypothesis is used to set the critical value of the test statistic. Under the null and alternative hypotheses (with $\hat{\mu} = \bar{y}$ under Model 2),

$$L_1 = \frac{1}{(\sqrt{2\pi}\sigma)^n} \exp\left\{-\frac{1}{2\sigma^2}[RSS + n(\bar{y} - \mu_1)^2]\right\}$$

$$L_2(\mu) = \frac{1}{(\sqrt{2\pi}\sigma)^n} \exp\left\{-\frac{1}{2\sigma^2}[RSS + n(\bar{y} - \mu)^2]\right\}$$

$$L_2(\hat{\mu}) = \frac{1}{(\sqrt{2\pi}\sigma)^n} \exp\left\{-\frac{1}{2\sigma^2}[RSS]\right\}$$

$$ML\,R_{12} = L_1/L_2(\hat{\mu})$$

$$= \exp\left\{-\frac{1}{2}\frac{n(\bar{y}-\mu_1)^2}{\sigma^2}\right\}$$

$$-2\log ML\,R_{12} = \frac{n(\bar{y}-\mu_1)^2}{\sigma^2}$$

which has a χ_1^2 sampling distribution under the null hypothesis.

The null hypothesis is rejected for small values of $ML\,R_{12}$, which is equivalent to large values of $-2\log RML_{12}$. For a test of size 5% the null hypothesis is rejected when $z_1^2 = n(\bar{y}-\mu_1)^2/\sigma^2 > 3.84$. Here $z_1 = 2$, so the hypothesis is (just) rejected, with a p-value of 0.0456. Correspondingly, the frequentist 95% confidence interval for μ, which is $\mu \in \bar{y} \pm 1.96\sigma/\sqrt{n} = [0.008, 0.792]$, (just) excludes $\mu = 0$.

2.6.2 Sequential sampling approach

The sequential sampling approach has an immediate difficulty. We need the size and power of the test of the null against the alternative, but the alternative is not specified. To apply the sequential test we have to find a way to specify the alternative and the desired power against it. Wald (1947) used a *weight function* $w(\mu)$ for this purpose. He considered the power function $P(\mu)$ of the test as a function of the alternative mean μ, and took an *average* of this function with respect to the weight function $w(\mu)$:

$$\bar{P} = \int P(\mu)w(\mu)d\mu.$$

If the weight function is regarded as a *prior distribution* on μ, this leads to a form of sequential analysis closely related to a form of Bayes analysis, as discussed in detail in Berger et al. (1997), and by their invited discussants. We do not discuss this further, and now describe our approach.

2.7 Posterior likelihood approach

Before giving the usual Bayesian analysis, we describe our approach as set out in Aitkin (1997), following Dempster (1974, 1997). As before we compute the likelihood ratio, which now depends on μ:

$$LR_{12}(\mu) = L(\mu_1)/L(\mu)$$

$$= \exp\left\{-\frac{n}{2\sigma^2}[(\bar{y}-\mu_1)^2 - (\bar{y}-\mu)^2)]\right\}.$$

As before, we express the likelihood ratio in terms of z-statistics under each model. We defined $z_1 = \sqrt{n}(\bar{y}-\mu_1)/\sigma$ above, and define $Z(\mu) = \sqrt{n}(\bar{y}-\mu)/\sigma$; then

$$LR_{12}(\mu) = \exp\left\{-\frac{1}{2}[z_1^2 - Z^2(\mu)]\right\}.$$

We define the *deviances* D_1 and $D_2(\mu)$ for Models 1 and 2 by

$$D_1 = -2\log L(\mu_1), \quad D_2(\mu) = -2\log L(\mu),$$

and the *deviance difference* between Models 1 and 2 by

$$D_{12}(\mu) = -2\log LR_{12}(\mu) = z_1^2 - Z^2(\mu).$$

In frequentist analysis, the term *deviance* is usually used for the statistic $-2\log L(\hat{\theta})$ where $\hat{\theta}$ is the maximum likelihood estimate of θ, or for the difference between this value and the value for the "saturated" model with a parameter for each observation. To avoid confusion we will refer to the frequentist statistic $-2\log L(\hat{\theta})$ as the *frequentist deviance*, and reserve the unqualified term "deviance" for the Bayesian form $-2\log L(\theta)$.

Given the data $\mathbf{y}$, the value of z_1 is known, but $Z(\mu)$ is unknown because it depends on the unknown μ. However, given a prior distribution $\pi(\mu)$, we have immediately the posterior distribution $\pi(\mu \mid \mathbf{y})$, and this gives the corresponding posterior distribution of $Z(\mu)$.

For example, if μ has a flat prior distribution over a large range beyond the limits of the data, its limiting posterior distribution as the range goes to infinity will be normal $N(\bar{y}, \sigma^2/n)$. (So a 95% symmetric credible interval for μ is $\bar{y}\pm 1.96\sigma/\sqrt{n}$, which is identical to the frequentist 95% confidence interval.)

Then the posterior distribution of $Z(\mu)$ will be $N(0, 1)$, so $Z^2(\mu)$ has a posterior χ_1^2 distribution, and the posterior distribution of $D_{12}(\mu)$ is *a location-shifted negative χ_1^2 distribution*. The amount of shift is the *squared frequentist z-statistic for testing the null hypothesis*. We give a general large-sample result in Section 2.7.1.

Our criterion for model choice is the value of the likelihood ratio. Without a specified alternative, the best we can do is to make *posterior probability* statements about μ, and transfer these to the *posterior distribution of the likelihood ratio*. It is simpler to give the posterior distribution of the deviance difference, as noted above. It follows directly that

$$\begin{aligned}
\Pr[LR > 1 \mid \mathbf{y}] &= \Pr[D_{12}(\mu) < 0 \mid \mathbf{y}] \\
&= \Pr[z_1^2 < \chi_1^2] \\
&= 2[1 - \Phi(z_1|)] \\
&= p,
\end{aligned}$$

the p-value of the observed $z_1 = 2$, which is 0.0456. That is (Dempster 1974, 1997) *the p-value is equal to the posterior probability that the likelihood ratio, for null hypothesis to alternative, is greater than 1.*

This result can be immediately transferred to the posterior probability of the null hypothesis: given equal prior probabilities on H_0 and H_1, the posterior probability of H_0 is $P[H_0 \mid \mathbf{y}] = LR/(1 + LR)$, so *the posterior probability is p that the posterior probability of H_0 is greater than 0.5*. More generally, if the prior odds on H_0 to H_1 is k, the posterior probability is p that the posterior odds on H_0 to H_1 is greater than k, so the posterior probability is p that the posterior probability of H_0 is greater than $k/(k+1)$.

This may seem convoluted: we would like to know the posterior probability of H_0, not just the posterior *distribution* of this probability. This is the price we have to pay for not knowing the value of μ under the alternative – we have a less precise answer (but one which is still a Bayesian posterior statement).

So the *p*-value in this example has a direct interpretation as a posterior probability: despite the integration over the sample space, which violates the principle of full conditioning on the observed data in Bayesian theory, the *p*-value *does* provide a measure of the strength of evidence against the null hypothesis. This *p*-value-posterior probability connection is not restricted to the normal model: Dempster (1997) and Aitkin (1997) showed that it extends generally to multiple-parameter simple null hypotheses, and to composite null hypotheses, about models with quadratic log-likelihoods and flat priors (that is, to most regular models in large samples), as we show in Section 2.7.1.

2.7.1 Large-sample result

The general result was given in Aitkin (1997, p. 258). We are given a large random sample $\mathbf{y}$ from the model $f(y \mid \psi)$, with $\psi' = (\theta', \phi')'$ where θ of dimension p is the parameter of interest, and ϕ of dimension q is the nuisance parameter. The composite null hypothesis is $H_1 : \theta = \theta_1$, the alternative is $H_2 : \theta \neq \theta_1$. The prior $\pi(\theta, \phi)$ is diffuse (flat) in the region of appreciable likelihood. We assume that the likelihood $L(\theta, \phi)$ has a maximum internal to the parameter space (though a maximum on the boundary does not invalidate our general approach, just the derivation of this large-sample result). We derive the posterior distribution of the deviance difference

$$-2 \log LR_{12} = -2 \log[L(\theta_1, \phi)/L(\theta, \phi)].$$

Let $(\widehat{\theta}, \widehat{\phi})$ be the MLE of (θ, ϕ), and $\widehat{\phi}_1$ be the MLE of ϕ given $\theta = \theta_1$. Expanding to the second-degree term the deviance D_1 about the MLE $\widehat{\phi}_1$ given θ_1, and the deviance D_2 about $(\widehat{\theta}, \widehat{\phi})$, we have

$$D_1(\theta_1, \phi) = D_1(\theta_1, \widehat{\phi}_1) + (\phi - \widehat{\phi}_1)' I_{22}(\phi - \widehat{\phi}_1)$$
$$D_2(\theta, \phi) = D_2(\widehat{\theta}, \widehat{\phi}) + (\psi - \widehat{\psi})' I (\psi - \widehat{\psi})$$

where I_{22} is the $q \times q$ submatrix of the information matrix I:

$$I_{22} = -\frac{\partial^2 \log L(\theta_1, \phi)}{\partial \phi \partial \phi'}$$

evaluated at $\phi = \widehat{\phi}_1 = \widehat{\phi} + I_{22}^{-1}I_{21}(\theta_0 - \widehat{\theta})$. Then

$$D_{12} = D_1(\theta_1, \widehat{\phi}_1) - D_2(\widehat{\theta}, \widehat{\phi})$$
$$+ (\phi - \widehat{\phi}_1)'I_{22}(\phi - \widehat{\phi}_1) - (\psi - \widehat{\psi})'I(\psi - \widehat{\psi}).$$

The first term is the frequentist likelihood ratio statistic, and the second term is the (negative) difference in "residual sums of squares" from "fitting" θ after ϕ. This has a χ_p^2 posterior distribution given the flat prior distribution. Thus, as for the simple null hypothesis case, *the deviance difference has a shifted negative χ_p^2 distribution, where the shift is the frequentist likelihood ratio test statistic.*

We do not need to *rely* on this result in small samples, because we always generate the *empirical posterior distribution* through posterior sampling. However, it is useful to examine the agreement between the empirical and asymptotic distributions in finite samples. We report this in some of the examples.

2.7.2 Strength of support for the null hypothesis

How strongly can the data *support* the null hypothesis? It is clear that the maximum support would come from $\bar{y} = \mu_1$, which would give a p-value of 1. In this case a p-value of 1 means that the posterior probability is 1 that the null hypothesis is better supported than the alternative, or equivalently that the posterior probability of the null hypothesis is greater than 0.5. This is certainly *support* for the null hypothesis, but it is far from convincing support. The reason is clear: values of μ close to μ_1 are nearly as well supported as μ_1 itself, so *there cannot be strong evidence in favor of a point null hypothesis against a general alternative hypothesis.*

This general result places our approach in conflict with the Bayes factor approach, discussed below, for which it is claimed that strong evidence for a point null hypothesis *can* be obtained. We discuss this conflict several times in later chapters.

2.7.3 Credible intervals for the deviance difference

In addition to the p-value information about the likelihood ratio or deviance difference, we may also use the posterior distribution to construct *credible intervals* for the true deviance difference. For example, if the 95% central credible interval for this difference include zero, then we do not have evidence from the data against the null hypothesis, in the sense that both hypotheses could be equally well supported.

The credible intervals are easily obtained analytically in this example: the 2.5 and 97.5 percentiles of χ_1^2 are 0.001 and 5.02, so the central 95% credible interval for the deviance difference is [4 − 5.02, 4 − 0.001], that is, [−1.02, 3.999]. This includes zero.

Small-sample results in general need simulation from the posterior distributions; Aitkin et al. (2005) gave some examples. For this example, we would make M random draws $\chi_1^{2[m]}$ from χ_1^2, subtract these from z_1^2 to give M

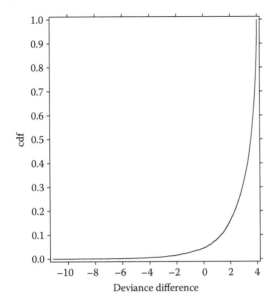

FIGURE 2.11
Deviance difference.

deviance difference draws $D_{12}^{[m]}$, and then exponentiate to give M likelihood ratio draws $LR_{12}^{[m]} = \exp\{-\frac{1}{2}D_{12}^{[m]}\}$. Figures 2.11 and 2.12 show the empirical cdf of $M = 10,000$ random draws of D_{12} and LR_{12} constructed in this way for the example with $\bar{y} = 0.4$. The likelihood ratio distribution plot is almost uninformative, and we will generally use the deviance difference plot for visual comparisons.

2.8 Bayes factors

In frequentist theory, estimation and testing are complementary, but in the Bayesian approach, the problems are completely different.... It may happen that conclusions based on estimation [posterior distribution] seem to contradict those from a Bayes factor. In this case the data seem unlikely under H_0, but if the Bayes factor turns out to be *in favor of H_0*, then the data are *even more unlikely* under H_1 than they would have been under H_0. Kass and Raftery (1995) [authors' emphasis].

* * *

... this book has little role for the non-Bayesian concept of hypothesis tests, especially where these relate to point null hypotheses of the form $\theta = \theta_0$ .... most of the difficulties in interpreting hypothesis tests arise from the

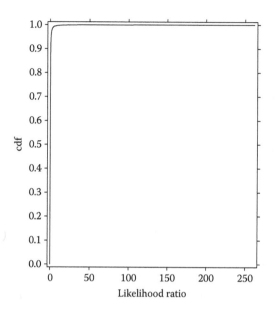

FIGURE 2.12
Likelihood ratio.

artificial dichotomy that is required between $\theta = \theta_0$ and $\theta \neq \theta_0$. Gelman et al. (2004, p. 250).

* * *

"...Bayes factors are rarely relevant in our approach to Bayesian statistics..." Gelman et al. (p. 192).

The reader may be surprised that we did not begin this section with the Bayes factor immediately; it is a common Bayesian tool for model comparisons (though as we note above many influential Bayesians dismiss it).

Under a parametric model M: $f_M(y \mid \theta)$ assuming high measurement precision δ, and given the value of θ, the probability of the observed data $\mathbf{y} = y_1, \ldots, y_n$ is

$$L_M(\theta) = \Pr[\mathbf{y} \mid \theta] = \prod_{i=1}^{n} f_M(y_i \mid \theta) \cdot \delta^n,$$

with prior distribution $\pi_M(\theta)$.

If we regard the likelihood as the *conditional* probability of $\mathbf{y}$ given θ and M, then the *unconditional* or *marginal* probability of the data given M is

$$\bar{L}_M = \int L_M(\theta) \pi_M(\theta) \mathrm{d}\theta$$

and this is called the *marginal* or *integrated* likelihood of the data.

Given two models M_1 and M_2 for the data, the Bayes factor is the *ratio of marginal likelihoods* under the two models:

$$BF_{12} = \bar{L}_1/\bar{L}_2.$$

It is then argued that the posterior probability of model M_1, given equal model prior probabilities, is $BF_{12}/(1 + BF_{12})$. The integration of the unknown parameters out of the likelihoods eliminates the uncertainty about the model parameters, and the model likelihoods, and the Bayes factor is interpreted as though the two models were *completely specified*. This is a simple and attractive approach, and is widely used by many Bayesians. However, the approach has a number of difficulties.

2.8.1 Difficulties with the Bayes factor

The first is logical – have we *really* eliminated the uncertainty about the model parameters, and their likelihoods, by integration? The integrated likelihood can also be described as the *prior mean* of the likelihood – it is the *expected value* of the likelihood with respect to our prior information. But what about the prior *variance* of the likelihood? This summarizes the *variation* of the likelihood (as a random function of θ) about its prior mean – the integrated likelihood. If the prior variance is large relative to the prior mean, we are misleading ourselves in thinking that the uncertainty in the likelihood has been eliminated – it has only been ignored.

The second difficulty is definitional. Any expectation with respect to the *prior* implies that the data have not yet been observed, in which case the function $\prod_i [f(y_i \mid \theta) \cdot \delta]$ is *not* the likelihood, but the joint probability of yet-to-be-observed *random variables* $Y_1, \ldots, Y_n$. So the "integrated likelihood" is the joint distribution of random variables drawn by a two-stage process, of drawing a random θ^* from this prior distribution, and a random sample $y_1, \ldots, y_n$ from $f(y \mid \theta^*)$. The marginal (compound) distribution of these random variables is not the same as the distribution of Y in the population from which the observed sample was drawn, and does not bear on the question of the value of θ in that population.

The third difficulty is both logical and computational. We cannot use an improper prior to compute the integrated likelihood, because an improper prior can be multiplied by an arbitrary constant, which will carry through to the integrated likelihood, making it arbitrary. So to compute an integrated likelihood *the prior must be proper*. This eliminates the usual improper noninformative priors widely used in posterior inference by most Bayesians.

Worse, any parameters in the priors (for example, in proper conjugate priors) will affect the value of the integrated likelihood, and this effect does not disappear with increasing sample size. This means that any user of a Bayes factor has to check the *sensitivity* of all the integrated likelihoods for the competing models to variations in the prior parameters. If the integrated likelihoods vary considerably over reasonable ranges in the prior parameters, it may not be possible to draw any firm conclusions about model preference from them.

A recent result illustrates a fundamental difficulty with the use of the integrated likelihood as a summary of the likelihood. If the two models being compared are *nested*, so that M_1 is a special case of M_2 under some parametric restriction, there is a simple relation between the Bayes factor and the posterior distribution of the likelihood ratio discussed in Section 2.7. Suppose that in the more general model M_2 the distribution depends on parameters θ and ϕ, while under the less general model M_1, θ is constrained to the value θ_1. (That is, model M_1 represents a *null hypothesis* that $\theta = \theta_1$.) We assume that the (proper) prior $\pi_1(\phi)$ under M_1 *corresponds* to the (proper) prior $\pi_2(\theta, \phi)$ under M_2, in the sense that $\pi_1(\phi) = \int \pi_2(\theta, \phi)d\theta$. (For example, if $\pi_2(\theta, \phi) = a(\theta)b(\phi)$, then $\pi_1(\phi) = b(\phi)$.)

Then

$$
\begin{aligned}
BF_{12} &= \frac{\bar{L}_1}{\bar{L}_2} \\
&= \frac{\int L_1(\theta_1, \phi)\pi_1(\phi)d\phi}{\int \int L_2(\theta, \phi)\pi_2(\theta, \phi)d\phi d\theta} \\
&= \frac{\int L_1(\theta_1, \phi)d\phi \int \pi_2(\theta, \phi)d\theta}{\int \int L_2(\theta, \phi)\pi_2(\theta, \phi)d\phi d\theta} \\
&= \frac{\int \int \frac{L_1(\theta_1, \phi)L_2(\theta, \phi)}{L_2(\theta, \phi)} \cdot \pi_2(\theta, \phi)d\phi d\theta}{\int \int L_2(\theta, \phi)\pi_2(\theta, \phi)d\phi d\theta} \\
&= \int \int \frac{L_1(\theta_1, \phi)}{L_2(\theta, \phi)} \cdot \frac{L_2(\theta, \phi)\pi_2(\theta, \phi)}{\int \int L_2(\theta, \phi)\pi_2(\theta, \phi)d\phi d\theta}d\phi d\theta \\
&= \int \int \frac{L_1(\theta_1, \phi)}{L_2(\theta, \phi)} \cdot \pi_2(\theta, \phi \mid y)d\phi d\theta \\
&= \int \int LR_{12}\pi_2(\theta, \phi \mid y)d\phi d\theta.
\end{aligned}
$$

That is, *the Bayes factor is equal to the posterior mean of the likelihood ratio between the models* (Kou et al. 2005; Nicolae et al. 2008).

A persistent criticism of the posterior likelihood approach, of the deviance information criterion (DIC) of Spiegelhalter et al. (2002), and of an earlier related use of the posterior mean of the likelihood (Aitkin 1991), has been based on the claim that these approaches are "using the data twice," or are "violating temporal coherence." The argument is that the data have already used up their information content in providing the likelihood to update the prior to the posterior, and cannot be used "again" to construct a posterior distribution for the likelihood itself – they are squeezed dry. (Some Bayesians have denied that the concept of the posterior distribution of the likelihood has any meaning.)

In the light of these criticisms, the Nicolae et al. (2008) result given above is remarkable. Though the Bayes factor is a ratio of *prior* means of the likelihoods, it is (for the nested case) also a *posterior* mean of the likelihood *ratio* between the nested models. For a posterior mean to be meaningful, one must accept that the posterior distribution itself is meaningful. If "using the data twice" invalidates the meaning or interpretation of the posterior distribution of the likelihood ratio, then it also invalidates the meaning or interpretation of the Bayes factor for nested models.

Assuming that we accept the validity of the posterior distribution of the likelihood ratio, we can repeat the difficulty with the Bayes factor already mentioned. It is the posterior *mean* of the likelihood ratio – but what about its posterior *variance* (discussed by Nicolae et al. 2008)? The posterior mean is a *one-point location* summary of a posterior distribution. But at least the *variability* in the posterior distribution – the *posterior variance* of the likelihood ratio – needs to be recognized. But even a two-point summary of a distribution is not enough – since we already have the full posterior distribution of the likelihood ratio, why not *use* this fully, as we described above?

Setting aside this issue for the moment, how well do Bayes factors work? The posterior mean of the likelihood ratio result is potentially *useful* because the direct computation of the Bayes factor through its definition requires *proper* priors for each model. This difficulty can apparently be avoided by using the posterior distribution of the alternative hypothesis model parameters.

However, the posterior mean of the likelihood ratio *may not be finite*. The simple normal model above provides an example of this. The likelihood ratio is

$$L R_{12}(\mu) = \exp\left\{-\frac{1}{2}[z_1^2 - Z^2(\mu)]\right\}.$$

and its posterior expectation is

$$E[L R_{12}(\mu)] = \exp\left\{-\frac{1}{2}z_1^2\right\} \cdot E\left[\exp\left\{\frac{1}{2}Z^2(\mu)\right\}\right].$$

The second term in the product is the moment-generating function $M(\theta)$ of $Z^2(\mu) \sim \chi_1^2$ evaluated at $\theta = 1/2$. The moment-generating function of χ_ν^2 is $(1 - 2\theta)^{-\nu/2}$, which is, for $\theta = 1/2$ and $\nu = 1$, $(1 - 1)^{-1/2}$ and is infinite.

This difficulty was predictable from our previous discussion, but it afflicts the Bayes factor in all the normal regression models. The problem is not solved by using a proper flat prior for μ on a finite interval, because $E[L R_{12}(\mu)] \to \infty$ as the prior interval width increases. To avoid it, Bayes factors have to use informative priors which do not give, in the likelihood ratio, a term involving e^{Z^2}. (We give an example in Section 2.8.2.) For this reason Bayes factors are not commonly used for the normal regression family of models.

Bayesians are divided in their views of Bayes factors, as noted above: Kass and Raftery argue strongly for them, while Gelman, Carlin, Stern, and Rubin are generally dismissive of them.

However, the difficulties with the Bayes factor do not prevent us from evaluating the posterior distribution of the likelihood ratio or deviance difference in general models, and we follow this approach throughout the book.

2.8.2 Conjugate prior difficulties

The above problem with Bayes factors can be lessened, but not avoided, by using an informative prior. For example, if the prior for μ under H_2 is taken to be $N(\mu_2, \sigma^2/m)$, the Bayes factor is

$$BF_{12} = \exp\left\{-\frac{1}{2}z_1^2\right\} \cdot \sqrt{(n+m)/m}\,\exp\left[\frac{m}{2(n+m)}z_2^2\right],$$

where $z_2 = \sqrt{n}(\bar{y} - \mu_2)/\sigma$ measures the discrepancy between the data and the prior. This discrepancy will be upweighted for small m by the factor $\sqrt{(n+m)/m}$ (the exponential term in z_2 will be close to 1) and will give support to the null hypothesis, which may outweigh the evidence from z_1. The posterior probability that the likelihood ratio $L_{12} > 1$ is

$$\Pr[LR_{12} > 1] = 2\left[1 - \Phi\left(\frac{z_1 - mz_2/(m+n)}{\sqrt{n/(n+m)}}\right)\right].$$

Here z_2 has a small effect which is downweighted for small m.

As $\mu_2 \to \mu_1$, $z_2 \to z_1$, and the Bayes factor approaches

$$BF_{12} = \sqrt{\frac{n+m}{m}} \cdot \exp\left\{-\frac{nz_1^2}{2(n+m)}\right\},$$

while the posterior probability that $LR > 1$ approaches

$$\Pr[LR_{12} > 1] = 2\left[1 - \Phi\left(\sqrt{\frac{n}{m+n}}z_1\right)\right].$$

The factor $\sqrt{(n+m)/m}$ will still affect the Bayes factor in the same way for small m, while the factor $\sqrt{n/(m+n)}$ will have a negligible effect on the posterior probability for small m.

As $m \to 0$, the Bayes factor approaches infinity, while the posterior probability that $LR > 1$ approaches

$$\Pr[LR_{12} > 1] = 1 - \Phi(z_1).$$

So the Bayes factor is very sensitive to both the specified prior mean μ_2 and the "prior sample size" m and diverges as the prior becomes diffuse. The posterior distribution of the likelihood ratio is very little affected by the prior mean μ_2 and is unaffected by the diffuse limit for m.

The claim that a null hypothesis can be strongly supported follows from the divergence of the Bayes factor to infinity as the prior becomes diffuse relative to the likelihood: the posterior probability of the null hypothesis tends to 1. Since this occurs *regardless of the observed data*, it cannot be taken as *data support* for the null hypothesis; rather, it is *prior* support for the null hypothesis, resulting from the diffuse prior on the alternative overpowering the likelihood from the data – *any* data.

This property of the Bayes factor has been known since the Lindley/Bartlett paradox papers of 1957 (Lindley 1957; Bartlett 1957). Despite the simplicity of the issue, and the reservations expressed by some Bayesians (notably, Gelman et al., 2004), the use and recommendation of Bayes factors is steadily increasing. A recent example due to Stone (1997, in the discussion of Aitkin 1997) shows how widespread is the confusion.

2.8.3 Stone example

The example from Stone (1997) is of a physicist running a particle-counting experiment who wishes to identify the proportion θ of a certain type of particle. He has a well-defined scientific (null) hypothesis H_1 that $\theta = 0.2$ precisely. There is no specific alternative. He counts $n = 527, 135$ particles and finds $r = 106, 298$ of the specified type. What is the strength of the evidence against H_1?

The binomial likelihood function

$$L(\theta) = \binom{n}{r}\theta^r(1-\theta)^{n-r} \doteq L(\hat{\theta})\exp\left\{-\frac{(\theta-\hat{\theta})^2}{2SE(\hat{\theta})^2}\right\}$$

is maximized at $\theta = \hat{\theta} = 0.201652$ with standard error $SE(\hat{\theta}) = 0.0005526$. The standardized departure from the null hypothesis is

$$Z_1 = (\theta_1 - \hat{\theta})/SE(\hat{\theta}) = 0.001652/0.0005526 = 2.9895,$$

with a two-sided p-value of 0.0028. The maximized likelihood ratio is $L(\theta_1)/L(\hat{\theta}) = 0.01146$.

The physicist uses the uniform prior $\pi(\theta) = 1$ on $0 < \theta < 1$ under the alternative hypothesis, and computes the Bayes factor

$$B = L(\theta_1)/\int_0^1 L(\theta)\pi(\theta)d\theta.$$

The denominator is

$$\bar{L} = \binom{n}{r}\int_0^1\theta^r(1-\theta)^{n-r}d\theta$$

$$= \binom{n}{r}B(r+1, n-r+1)$$

$$\doteq L(\hat{\theta})\int_0^1\exp\left\{-\frac{(\theta-\hat{\theta})^2}{2SE(\hat{\theta})^2}\right\}d\theta$$

$$= \sqrt{2\pi}SE(\hat{\theta})L(\hat{\theta})$$

$$= L(\hat{\theta})/f(\hat{\theta}).$$

Here $f(\hat{\theta})$ is the normal posterior density $N(\hat{\theta}, SE(\hat{\theta})^2)$ of θ evaluated at the mean $\hat{\theta}$; since the sample size is so large the actual posterior beta density is very closely normal.

The Bayes factor is thus

$$BF_{12} = L(\theta_1)/\bar{L}$$
$$= f(\hat{\theta}) \cdot L(\theta_1)/L(\hat{\theta}),$$

a simple multiple of the maximized likelihood ratio. The multiplier is

$$f(\hat{\theta}) = \frac{1}{\sqrt{2\pi}\,SE(\hat{\theta})} = \frac{1}{0.0013851} = 721.937,$$

giving the Bayes factor

$$B_{12} = 721.937 \cdot 0.01146 = 8.27,$$

which appears to be quite strong evidence in favor of the null hypothesis.

Thus the *p*-value and Bayes factor are in clear conflict. However, the posterior distribution of θ is *not* in conflict with the *p*-value, since the posterior probability that $\theta > 0.2$ is $\Phi(2.990) = 0.9986 = 1 - p/2$. Any Bayesian using the uniform prior must have a very strong posterior belief that the true value of θ is larger than 0.2. Equivalently, the 99% HPD interval for θ is

$$\theta \in \hat{\theta} \pm 2.576 SE(\hat{\theta}) = (0.20023, 0.20308)$$

which is identical to the 99% confidence interval, and excludes θ_1.

So the use of the Bayes factor leads to an evidential conclusion about the null hypothesis which is inconsistent with that from the posterior distribution using the same prior. It is a general feature of the "Lindley paradox" or "Bartlett paradox" that this inconsistency is most clearly visible in very large samples (Aitkin, 1991; 1997). This inconsistency is not a conflict between Bayes axioms and frequentist axioms, but is within Bayesian theory, between two conflicting uses of the likelihood function. With increasing sample size the likelihood becomes increasingly sharply peaked, and so the integration of this increasingly sharply-peaked function with respect to a uniform weight function over a fixed range produces decreasingly small integrated likelihoods. In this example with its gigantic sample size, the maximized likelihood is reduced to the integrated likelihood by the factor 721.94, giving a minute integrated likelihood against which the very small value of $L(\theta_1)$ looks large.

The difficulty in interpreting evidence through the Bayes factor in this example is clear if we note that it is just a scale multiple of the maximized likelihood ratio, and so any value of θ in the region of $\hat{\theta}$ will have a *very* large Bayes factor, providing very much stronger evidence than a Bayes factor of 8 in support of the value 0.2: at the MLE $\theta_1 = \hat{\theta}$, the Bayes factor takes the value 721.94. Is 8 persuasive compared with this?

2.9 The comparison of unrelated models

We consider finally the comparison of unrelated models, neither of which can be expressed as a submodel of the other. This is a particularly difficult problem for the frequentist theory, with few optimal results (Pace and Salvan 1990). The general approach can be seen from the comparison of the two models.

Under Model 1, Y has model $f_1(y \mid \theta_1)$ with likelihood $L_1(\theta_1)$ and prior $\pi_1(\theta_1)$, while under Model 2 Y has model $f_2(y \mid \theta_2)$ with likelihood $L_2(\theta_2)$ and prior $\pi_2(\theta_2)$. If θ_1 and θ_2 were known (as θ_{1T} and θ_{2T}), we would compute the likelihood ratio $LR_{12} = L_1(\theta_{1T})/L_2(\theta_{2T})$.

We make M draws $\theta_1^{[m]}$ from the posterior $\pi_1(\theta_1 \mid \mathbf{y})$ and substitute them into the Model 1 likelihood to give M draws $L_1^{[m]} = L_1(\theta_1^{[m]})$ from L_1, and make M *independent* draws $\theta_2^{[m]}$ from the posterior $\pi_2(\theta_2 \mid \mathbf{y})$ and substitute them into the Model 2 likelihood to give M draws $L_2^{[m]} = L_2(\theta_2^{[m]})$ from L_2. Then the M values $LR_{12}^{[m]} = L_1^{[m]}/L_2^{[m]}$ are M random draws from the posterior distribution of L_1/L_2. We evaluate these more simply through the corresponding deviances, $D_1^{[m]} = -2 \log L_1^{[m]}$ and $D_2^{[m]} = -2 \log L_2^{[m]}$, and their difference, $D_{12}^{[m]} = D_1^{[m]} - D_2^{[m]}$.

The independence of the parameter draws is important: if the same random seed is used for both sets of draws, the likelihood draws will be correlated and their ratio distribution will be incorrect. This possibility can always be avoided by *randomly permuting* one set of draws relative to the other before pairing them. Different permutations will give slightly different posterior distributions of the ratio: this is an inherent feature of simulation results. The effect is reduced by increasing the simulation sample size.

In more general models requiring Markov chain Monte Carlo methods for the parameter posteriors, this approach has been used by Fox (2005), and by Congdon (2005, 2006a, 2006b) under the name *parallel chains*, though it is not necessary to run the chains in parallel for the different models.

2.9.1 Large-sample result

For regular models $f(y \mid \theta)$ with flat priors, using the second-order Taylor expansion of the deviance about $\hat{\theta}$ gives:

$$-2 \log L(\theta) \doteq -2 \log L(\hat{\theta}) - 2(\theta - \hat{\theta})' \ell'(\hat{\theta}) - (\theta - \hat{\theta})' \ell''(\hat{\theta})(\theta - \hat{\theta})$$

$$= -2 \log L(\hat{\theta}) + (\theta - \hat{\theta})' I(\hat{\theta})(\theta - \hat{\theta})$$

$$L(\theta) \doteq L(\hat{\theta}) \cdot \exp[-(\theta - \hat{\theta})' I(\hat{\theta})(\theta - \hat{\theta})/2]$$

$$\pi(\theta \mid \mathbf{y}) \doteq c \cdot \exp[-(\theta - \hat{\theta})' I(\hat{\theta})(\theta - \hat{\theta})/2].$$

So asymptotically (and loosely in the first result)

$$\theta \mid \mathbf{y} \sim N(\hat{\theta}, I(\hat{\theta})^{-1}),$$

$$(\theta - \hat{\theta})' I(\hat{\theta})(\theta - \hat{\theta}) \mid \mathbf{y} \sim \chi^2_{p},$$

$$-2 \log L(\theta) \sim -2 \log L(\hat{\theta}) + \chi^2_{p}.$$

So the deviance $-2 \log L(\theta)$ has an asymptotic *shifted* χ_p^2 distribution, shifted by the *frequentist deviance* $-2 \log L(\hat{\theta})$, where p is the dimension of θ.

This is a simple and remarkable result (given by a different argument in Bickel and Ghosh (1990)): for large samples from regular models with noninformative priors, *the posterior distribution of the likelihood*, or of the deviance, depends on the data *only through the maximized likelihood*. So in large samples, the comparison of nonnested regular models depends only on their maximized likelihood ratio, or their *frequentist deviance difference*. The *calibration* of the frequentist deviance difference depends on the distribution of the difference between two independent χ^2 variables, which has no exact distribution, but is very easily simulated.

Further, the agreement between the asymptotic and the empirical distributions is very easily assessed, by simply plotting the empirical and asymptotic cdfs in the same graph. (This contrasts strongly with the asymptotic repeated-sampling χ^2 distribution of the frequentist likelihood ratio test statistic, for which there is no validation possible from the observed data except through a Bartlett correction, which may not be sufficient in small samples.) Our experience with heavily parametrized models (for example, the normal mixture distributions in Chapter 8) shows that with increasing parameter dimension, the *achieved* maximum in 10,000 draws from the posterior distribution of the likelihood is increasingly far below the analytic maximum.

However, we do not *need* to rely on the asymptotic distribution, as the empirical distribution provides all the necessary information (subject only to the sampling variation in proportions computed from a simulation sample size of 10,000).

One further important point is that if the models are actually nested, with p_1 parameters under the null Model 1 and $p_2 > p_1$ under the alternative Model 2, then by drawing from the posterior distribution of the common parameters under the alternative, we obtain the nested result, with the χ^2 distributions generated from the *same* parameter draws. The distribution of the deviance difference, instead of being the difference between $\chi_{p_2}^2$ and $\chi_{p_1}^2$, is $\chi_{p_2 - p_1}^2$, which has the same mean but a much smaller variance: $2(p_2 - p_1)$ instead of $2(p_2 + p_1)$.

2.9.2 Bayes factor

An elegant result developed in detail in Newton and Raftery (1994) appears to avoid the prior integral problem for nonnested models. They expressed the prior mean of the likelihood

$$\bar{L} = \int_\theta L(\theta)\pi(\theta)d\theta$$

as a *posterior expectation*: from Bayes's theorem

$$\pi(\theta \mid \mathbf{y}) = L(\theta)\pi(\theta)/\bar{L},$$

we have

$$\frac{\pi(\theta)}{\bar{L}} = \frac{\pi(\theta \mid \mathbf{y})}{L(\theta)},$$

and on taking expectations over θ, if the prior is proper,

$$\frac{1}{\bar{L}} = E\left[\frac{1}{L(\theta)} \mid \mathbf{y}\right]:$$

the reciprocal of the integrated likelihood equals the posterior mean of the reciprocal of the likelihood. This is known as the *harmonic mean identity.*

Unfortunately, this has the same problem as for the nested case. Using the asymptotic expansion from the previous section, we have

$$\begin{aligned}
\frac{1}{\bar{L}} &= \frac{E[e^{\frac{1}{2}\chi_p^2}]}{L(\hat{\theta})} \\
&= \frac{M_p(1/2)}{L(\hat{\theta})} \\
&= \frac{(1 - 2 \cdot (1/2))^{-p/2}}{L(\hat{\theta})} \\
&= \frac{0^{-p/2}}{L(\hat{\theta})} \\
&= \infty
\end{aligned}$$

and $\bar{L}$ is not defined.

Newton and Raftery (1994) used a conjugate prior for θ to avoid this difficulty. This prior *has to be informative* to give a finite posterior mean, but the posterior *variance* (and higher moments) of $1/L(\theta)$ approach infinity as the prior becomes more diffuse. They used M random draws $\theta^{[m]}$ from the posterior of θ substituted in $1/L(\theta)$ to give draws $1/L(\theta^{[m]})$.

The "estimate"

$$\frac{\tilde{1}}{\bar{L}} = \frac{\sum_{m=1}^{M} 1/L(\theta^{[m]})}{M}$$

of $1/L$ has very large variance, and so is close to *inconsistency* – it may not converge to the true mean as the simulation size $M \to \infty$. So this approach does not provide a satisfactory analysis.

2.9.3 Example

We illustrate the posterior likelihood ratio analysis with a well-known example, of comparing a Poisson with a geometric distribution (Cox; 1961, 1962). The data are $n = 30$ event counts y_i from either a Poisson or a geometric distribution, and are given in Table 2.2 as frequencies f.

TABLE 2.2

Counts from Cox (1961)

y	0	1	2	3	> 3
f	12	11	6	1	0

The Poisson and geometric likelihoods and deviances (parametrized in terms of the mean μ) are

$$P(\mu) = \prod_i \exp(-\mu)\mu^{y_i}/y_i!$$

$$= \exp(-n\mu)\mu^T/P$$

$$D_P(\mu) = 2[n\mu - T\log\mu + \log P]$$

$$G(\mu) = \prod_i \left(\frac{\mu}{1+\mu}\right)^{y_i} \frac{1}{1+\mu}$$

$$= \frac{\mu^T}{(1+\mu)^{T+n}}$$

$$D_G(\mu) = 2[(T+n)\log(1+\mu) - T\log\mu]$$

where $T = \sum_i y_i = 26$, $P = \prod_i y_i! = 384$. The two likelihoods are shown in Figure 2.13 (Poisson solid curve, geometric dot-dashed).

The ratio of maximized likelihoods (at the common MLE of $\bar{y}$) is 20.13 in favour of the Poisson. This corresponds to a frequentist deviance difference of 6.00. For any value of μ other than the MLE, the likelihood ratio between the models is smaller than 20.13.

We take the same flat prior on μ under each model; the posterior distributions of μ are then gamma $(T+1, n)$ for the Poisson model and beta $(T+1, n-1)$ for the geometric. The cdfs of $M = 10,000$ draws from the posterior distributions of the Poisson (solid) and geometric (dot-dashed) deviances are shown in Figure 2.14, together with the asymptotic χ_1^2 distribution shifted by the frequentist deviances (dotted).

The Poisson clearly dominates the geometric. Both deviance distributions are very well represented by the shifted χ_1^2 distribution. Figure 2.15 shows the deviance difference; the median is 6.01 – almost the same as the frequentist deviance difference – and the central 95% credible interval for the true deviance difference is [1.58, 10.64]. This translates to a 95% credible interval for the posterior probability of the Poisson model of [0.688, 0.995]; the median value is 0.952. The credible interval is heavily skewed.

The evidence in favor of the Poisson is quite strong, though not as strong as the ratio of maximized likelihoods suggests, because of the long tail of the posterior probability distribution.

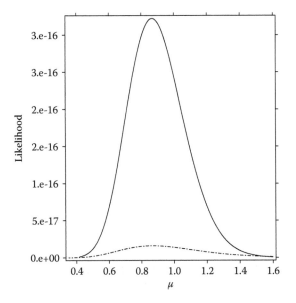

FIGURE 2.13
Poisson and geometric likelihoods.

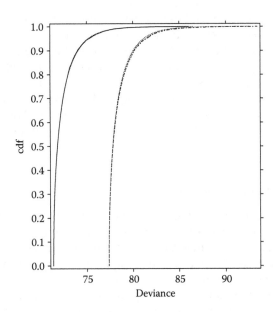

FIGURE 2.14
Poisson and geometric deviances.

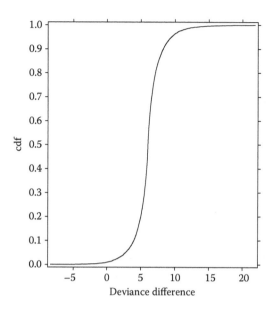

FIGURE 2.15
Deviance difference distribution.

A Bayes factor can be calculated here by the device of using the same finite uniform distribution for μ on $[a, b]$, taking the ratio of integrated likelihoods, in which the factor $(b - a)$ cancels, and then letting $(b - a) \rightarrow \infty$. The resulting Bayes factor is 14.22 with $2\log(BF) = 5.31$, near the median of the deviance difference. The posterior probability of the Poisson model using this value for the likelihood ratio is 0.934. If the prior $d\mu/\mu$ is used instead of $d\mu$, the Bayes factor becomes 14.71, with $2\log(BF) = 5.38$, and the 95% credible interval for the deviance difference becomes [1.62, 10.62]. The posterior probability of the Poisson model using this value for the likelihood ratio is 0.936.

2.9.4   Misleading conclusions from Bayes factors

Misleading conclusions from Bayes factors are not restricted to point null hypotheses. They occur quite generally in comparisons of nonnested models, when the parameter spaces over which the integrations are carried out have different effects in the different models. A striking example, appearing in a paper by Lee (2004), is discussed in Liu and Aitkin (2008).

Squire (1989) examined how long-term recognition memory for names of television shows depended on the time in years when the show was last broadcast. Let $y_i (i = 1, \ldots, 15)$ denote recognition accuracy, measured in average proportion correct, and t_i denote the time intervals, as a proportion of 15 years. Each model is indexed by two parameters, m and b, relating the mean accuracy μ_i to the time interval t_i. The five models were

- Exponential: $\mu_i = b \exp(-mt_i)$
- Linear: $\mu_i = b - mt_i$
- Hyperbolic: $\mu_i = 1/(mt_i + b)$
- Logarithmic: $\mu_i = b - m \ln t_i$
- Power: $\mu_i = bt_i^{-m}$

For each model, the observed proportions y_i were assumed to arise from a normal sampling distribution with mean μ_i, and common variance σ^2 for all time intervals. Given the sample standard deviations reported in Squire (1989), Lee (2004) argued that a reasonable estimate of σ was 0.25. Importantly, his conclusions regarding model comparisons were robust across a broad range of estimates for σ.

The models were compared by Bayes factors, and uniform priors were assigned to the parameters m and b on the interval [0, 2]. Under this prior specification, the hyperbolic model had the highest integrated likelihood, followed by the exponential, linear, power, and logarithmic models. The Bayes factors for the hyperbolic compared to the other models were 3.4, 6.7, 13.0, and 16.4, respectively. Lee (p. 315) concluded that

> ... at the estimated level of data precision, the hyperbolic model constitutes the best balance between fit and inherent complexity, and is most strongly supported by Squire's (1989) data.

The common parameter space for all models *appears* to allow a fair comparison across the integrated likelihoods. However, inspection of the likelihood contours (Figures 2.16 through 2.20) show that these have very different dispersions: the models with very tight and well-defined likelihood contours have the highest maximized likelihoods, but the smallest integrated likelihoods because of their tight likelihood contours. (The unshaded areas in the figures show the legitimate range of parameter values giving positive predicted values.)

This hazard may occur in all such nonnested comparisons, but is avoided completely by the comparison of likelihood or deviance distributions for each model (also given in Liu and Aitkin 2008). Figure 2.21 shows the deviance distributions, computed from independent samples of 10,000.

The stochastic order of model deviances is quite different, from best to worst: log, power, hyperbolic, exponential, linear. Table 2.3 gives the medians and the central 95% credible intervals for the differences between the log and the other model deviances.

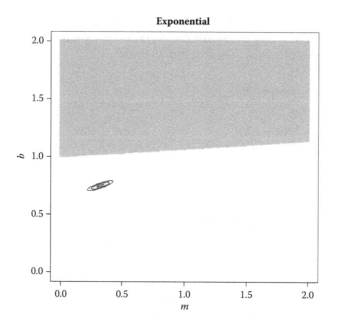

FIGURE 2.16

Parameter space for the exponential model. The shaded region represents parameter values that generate impossible predictions for proportion of recall (i.e., less than 0, or greater than 1). The contours in the unshaded region represent the posterior distribution.

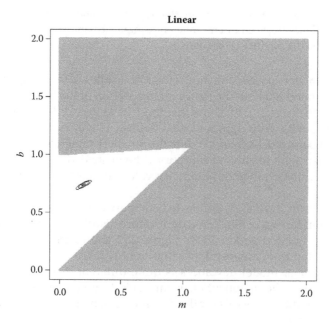

FIGURE 2.17

Parameter space for the linear model.

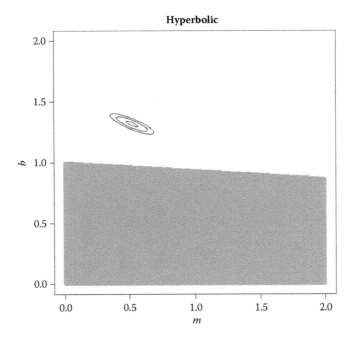

FIGURE 2.18
Parameter space for the hyperbolic model.

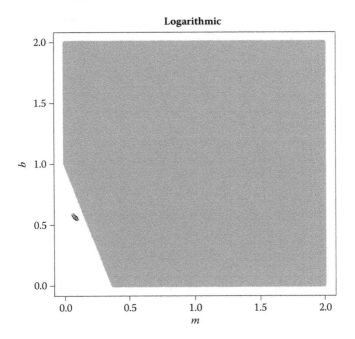

FIGURE 2.19
Parameter space for the logarithmic model.

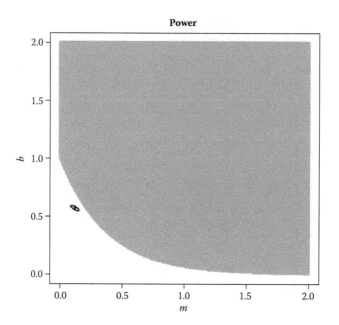

FIGURE 2.20
Parameter space for the power model.

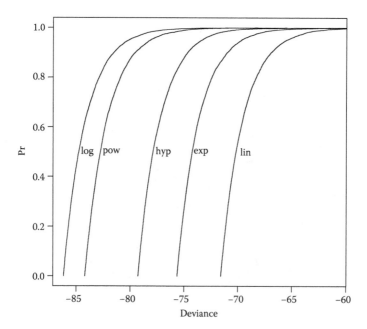

FIGURE 2.21
Posterior deviance distributions for retention functions (log = logarithmic, pow = power, hyp = hyperbolic, exp = exponential, lin = linear). Larger deviance values indicate worse fit.

TABLE 2.3

Credible Intervals of 95% for Deviance Difference

Function	2.5%	50%	97.5%
Power	−4.0	2.0	8.0
Hyperbolic	0.8	6.9	12.7
Exponential	4.6	10.5	16.5
Linear	8.7	14.6	20.6

Note: The deviance differences are with respect to the logarithmic model. Positive values indicate evidence
in favor of the logarithmic model. Negative values
indicate evidence in favor of the alternative model.

Only the power model is not inferior to the log model (the credible interval
for the deviance difference includes zero).

This inversion of order relative to the integrated likelihoods was not peculiar to the deviance distributions; it occurred for *all* the other methods for
comparing models, frequentist or Bayesian. The preference ordering for all
these other methods was the same as that given by their maximized likelihoods, as would be expected from the asymptotic distribution result.

2.9.5 Modified Bayes factors

These difficulties of the Bayes factor have led to several proposals for variations to it to avoid them. These adopt the approach of cross-validation procedures, by splitting the sample data into a minimal "training set," which
updates the improper flat prior to a minimally informative proper posterior,
and then computing a Bayes factor from the remaining data subset.

O'Hagan (1995) and Berger and Pericchi (1996) proposed different procedures: the "fractional Bayes factor" (FBF) and the "intrinsic Bayes factor"
(IBF), respectively, for this purpose. The FBF assumes arbitrarily that the training sample likelihood is a power of the full sample likelihood, and the likelihood for the remaining data set is the complementary fractional power of
the full sample likelihood. The IBF does not assume this, but averages all the
Bayes factors (additively or multiplicatively) over all possible assignments of
the sample observations to the training set. These procedures both have ad
hoc aspects and have not been widely adopted.

Spiegelhalter et al. (2002) proposed the deviance information criterion, using the posterior mean of the deviance penalized by a function of the number
of model parameters. These proposals were stimulated by an earlier proposal
of mine (Aitkin 1991) to replace the prior integrals of the likelihoods in the
Bayes factor by the posterior integrals, leading to the use of the posterior
means of the likelihoods in the same way as the prior means in the Bayes
factor, giving what I termed the posterior Bayes factor.

This idea was heavily criticized, on the grounds mentioned above, of "using the data twice" and "violating temporal coherence." Whatever the merits of the approach, its weakness in my view is shared by the Akaike information criterion (AIC), Bayesian information criterion (BIC), deviance information criterion (DIC), Bayes factor, fractional Bayes factor (FBF), and intrinsic Bayes factor (IBF) – the expression of evidence about the uncertain value of the likelihood by a single number. The difficulties suffered by all these approaches are remedied by the use of the full posterior distribution of the likelihood, or deviance.

In the remainder of this book, we apply this approach systematically to a wide range of standard and nonstandard modeling problems. We conclude this chapter with a simple example which illustrates some of the features of more complex examples discussed later in the book.

2.10   Example – GHQ score and psychiatric diagnosis

The data in Table 2.4 were published by Silvapulle (1981), and were discussed at length in Aitkin et al. (2009).

They come from a psychiatric study of the relation between psychiatric diagnosis (as case or noncase) and the value of the score on a 12-item General Health Questionnaire (GHQ), for 120 patients attending a general practitioner's surgery. Each patient was administered the GHQ, resulting in a score between 0 and 12, and was subsequently given a full psychiatric examination by a psychiatrist who did not know the patient's GHQ score. The patient was classified by the psychiatrist as either a "case," requiring psychiatric treatment, or a "noncase." The question of interest was whether the GHQ score, which could be obtained from the patient without the need for trained psychiatric staff, could indicate the need for psychiatric treatment. Specifically,

TABLE 2.4

GHQ Score for Cases and Noncases

GHQ	Men Cases	Men Noncases	Women Cases	Women Noncases	Total Cases	Total Noncases
0	0	18	2	42	2	60
1	0	8	2	14	2	22
2	1	1	4	5	5	6
3	0	0	3	1	3	1
4	1	0	2	1	3	1
5	3	0	3	0	6	0
6	0	0	1	0	1	0
7	2	0	1	0	3	0
8	0	0	3	0	3	0
9	0	0	1	0	1	0
10	1	0	0	0	1	0

given the value of GHQ score for a patient, what can be said about the probability that the patient is a psychiatric case? Sex of the patient is an additional variable.

Both men and women patients are heavily concentrated at the low end of the GHQ scale, where the overwhelming majority are classified as noncases. The small number of cases are spread over medium and high values of GHQ.

We model the relation between "caseness" and GHQ through a probit regression. Silvapulle (1981) and Aitkin et al. (2009) used a logit model: we choose the probit here to illustrate different possible Bayesian analyses. We ignore the sex classification which the analyses above found irrelevant.

For the data c_i, n_i, x_i of c_i cases and n_i noncases at GHQ score $x_i = i$, $i = 0, 1, \ldots 10$, we use the probit model

$$p_i = \Pr[case|x_i] = \Phi[\alpha + \beta x_i]$$

where Φ is the normal cdf. The likelihood is

$$L(\alpha, \beta) = \prod_{i=0}^{10} p_i^{c_i} (1 - p_i)^{n_i}$$

$$= \prod_{i=0}^{10} (\Phi[\alpha + \beta x_i])^{c_i} (1 - \Phi[\alpha + \beta x_i])^{n_i}.$$

The maximum likelihood estimates and (standard errors) are $\hat{\alpha} = -1.919$ (0.263), $\hat{\beta} = 0.788\,(0.144)$. The frequentist deviance is 1.93 with 9 degrees of freedom: the fit is very close. (The logit model fits equally well, with almost the same frequentist deviance.)

The correlation between $\hat{\alpha}$ and $\hat{\beta}$ in the likelihood is very high (-0.722); this could cause difficulties in Markov Chain Monte Carlo (MCMC) or in sampling directly from the normalized likelihood. Figure 2.22 shows the likelihood computed over a 100×100 grid (α_k, β_k), $k = 1, \ldots, 10,000$ centered on the MLEs.

We cannot orthogonalize the information matrix simply, because its terms are weighted sums of squares and cross-products with weights which depend on the parameter estimates. We can near-orthogonalize by (for example) simple *centering* – subtracting the mean of x (1.45) from x, which orthogonalizes the information matrix in the normal regression model. This gives the model

$$p_i = \Phi[\alpha^* + \beta(x_i - 1.45)]$$
$$= \Phi[\alpha^* - 1.45\beta + \beta x_i]$$

so that $\alpha^* - 1.45\beta = \alpha$.

Refitting the centered model by maximum likelihood, we find estimates and standard errors $\hat{\alpha}^* = -0.777\,(0.183)$, $\hat{\beta} = 0.788\,(0.144)$, with correlation 0.100. Computing the likelihood over a 100×100 grid (α_k^*, β_k), $k = 1, \ldots, 10,000$ around the centered estimates, we obtain Figure 2.23.

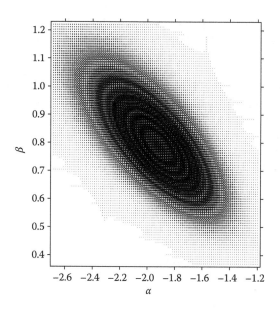

FIGURE 2.22
Likelihood in (α, β).

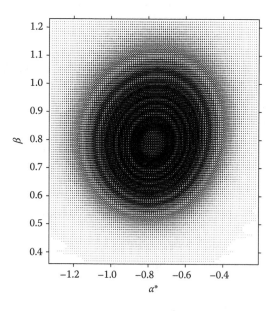

FIGURE 2.23
Likelihood in (α^*, β).

The reparametrization gives a more detailed computation of the likelihood over a much larger effective grid.

Our practical interest is in the prediction of caseness probability given the GHQ score. For example, at GHQ = 2, the observed proportion of cases is $5/11 = 0.455$, and the "fitted" proportion from the MLE is $\Phi[-0.343] = 0.366$. We need the posterior distribution of $\Phi[\alpha + 2\beta] = \Phi[\alpha^* + 0.55\beta]$, which is most simply computed by making $M = 10{,}000$ draws $(\alpha^{*[m]}, \beta^{[m]})$ from the joint posterior distribution of (α^*, β). Then we simply form the sums $\alpha^{*[m]} + 0.55\beta^{[m]}$ and compute the M normal cdf values $\Phi[\alpha^{*[m]} + 0.55\beta^{[m]}]$.

We need to specify prior distributions on α^* and β. We use flat priors, which leave the posterior distribution proportional to the likelihood. So we compute the posterior distribution by simply summing the likelihoods and normalizing by the sum, to give

$$\pi(\alpha_k^*, \beta_k \mid \mathbf{y}) = L(\alpha_k^*, \beta_k) / \sum_{\ell} L(\alpha_\ell^*, \beta_\ell).$$

We then draw a value α_k^*, β_k with probability $\pi(\alpha_k^*, \beta_k \mid \mathbf{y})$ and repeat the independent draws M times.

Figure 2.24 shows the cdfs of the case probabilities for all GHQ values from 0 to 9 (reading from the left side to the right).

The 95% credible intervals, and medians, from the posterior distributions for GHQ 0 to 6 are shown in Table 2.5 below together with the predicted

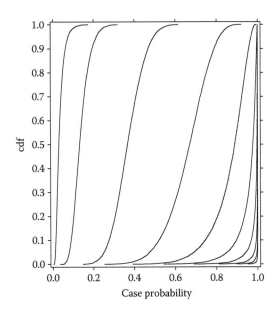

FIGURE 2.24
Case probabilities.

TABLE 2.5

Estimates and Intervals for Case Probability

GHQ	2.5%	Median	97.5%	2.5%	MLE	97.5%
0	0.009	0.029	0.090	0.009	0.031	0.077
1	0.072	0.132	0.234	0.055	0.118	0.208
2	0.230	0.367	0.520	0.212	0.360	0.552
3	0.436	0.670	0.836	0.467	0.704	0.891
4	0.641	0.888	0.976	0.701	0.909	0.983
5	0.804	0.977	0.999	0.857	0.977	0.999
6	0.912	0.997	1.000	0.935	0.994	0.999

values from the MLE, and approximate 95% confidence intervals based on the profile likelihoods, as given in Aitkin et al. (2009). The agreement is quite close though the approximate confidence intervals are narrower except at GHQ values 2 and 3: the maximization over the nuisance parameters overstates the information about them.

3

t-Tests and Normal Variance Tests

A remarkable feature of Bayesian analysis has been the absence, until very recently, of formal Bayes procedures paralleling the frequentist two-sample t-test – reliance has been placed instead on the credible interval for the mean difference. Bayes factors have not been used, for reasons we have already seen.

The one-sample test was discussed briefly in Aitkin (1997). In this chapter we give first a full exposition of the Bayesian analogues of the one- and two-sample t-tests (the latter with equal or unequal variances), the bread-and-butter of simple normal-based frequentist analysis. These sections are from unpublished papers by Aitkin (2006) and Aitkin and Liu (2007). We comment on the very recent Bayesian alternatives to the two-sample t-test.

The final sections deal with one- and two-sample tests for normal model variances. We resolve a long-standing argument over the role of the *marginal* or *restricted* likelihood for inference about the variance.

3.1 One-sample t-test

3.1.1 Credible interval

In the one-sample case, we observe data y_i, $i = 1, \ldots n$ from the model $Y \sim N(\mu, \sigma^2)$. We assume high precision in the measurement of y, and omit the measurement precision δ as it vanishes in the likelihood ratio. The likelihood function is

$$
\begin{aligned}
L(\mu, \sigma) &= \prod_{i=1}^{n} \frac{1}{\sqrt{2\pi}\sigma} \exp\left\{-\frac{1}{2}\frac{(y_i - \mu)^2}{\sigma^2}\right\} \\
&= c \cdot \frac{1}{\sigma^n} \exp\left\{-\frac{1}{2\sigma^2}[RSS + n(\bar{y} - \mu)^2]\right\} \\
&= c \cdot \frac{\sqrt{n}}{\sigma} \exp\left\{-\frac{1}{2}\frac{n(\bar{y} - \mu)^2}{\sigma^2}\right\} \cdot \frac{1}{\sigma^{n-1}} \exp\left\{-\frac{1}{2}\frac{RSS}{\sigma^2}\right\}
\end{aligned}
$$

where $RSS = \sum_i (y_i - \bar{y})^2$. A conjugate prior for μ, σ based on a (hypothetical) *prior sample of size m*, has the same form as the likelihood:

$$\pi(\mu, \sigma) = c' \cdot \frac{\sqrt{m}}{\sigma} \exp\left\{-\frac{1}{2}\frac{m(\mu_P - \mu)^2}{\sigma^2}\right\} \cdot \frac{1}{\sigma^v} \exp\left\{-\frac{1}{2}\frac{PSS}{\sigma^2}\right\},$$

where μ_P is the prior mean of μ and PSS is the "prior sum of squares" with v degrees of freedom defining the prior precision of σ. Thus μ has a normal prior distribution $N(\mu_P, \sigma^2/m)$ conditional on σ, and PSS/σ^2 has a marginal χ_v^2 prior distribution. The posterior distribution of μ given σ is $N([n\bar{y} + m\mu_P]/[n + m], \sigma^2/[n + m])$, while the marginal posterior distribution of $(RSS + PSS)/\sigma^2$ is χ_{n-1+v}^2. Letting m and $v \to 0$ (with $PSS = 0$), the diffuse limit of the conjugate prior is the improper prior $1/\sigma$, and the resulting posterior distribution of μ and σ^2 can be expressed as

$$\mu \mid \sigma, \mathbf{y} \sim N(\bar{y}, \sigma^2/n), \quad RSS/\sigma^2 \mid \mathbf{y} \sim \chi_{n-1}^2.$$

Integrating over σ, standard calculations show that the marginal posterior distribution of $t = \sqrt{n}(\mu - \bar{y})/s$ is t_{n-1}, where $s^2 = RSS/(n-1)$. The $100(1-\alpha)\%$ credible intervals for μ correspond exactly, for the diffuse prior, to the usual $100(1-\alpha)\%$ frequentist t confidence intervals. Here the posterior distribution is analytic, but we note for later use that the distribution of μ can be simulated in two stages:

- Make M random draws of RSS/σ^2 from its marginal χ_{n-1}^2 distribution, and hence M random draws $\sigma^{[m]}$ of σ.
- For each $\sigma^{[m]}$, make a random draw $\mu^{[m]}$ of μ from its conditional normal $N(\bar{y}, \sigma^{[m]2}/n)$ distribution.

The M draws $\mu^{[m]}$ ($m = 1, \ldots, M$) provide a random sample from the marginal posterior distribution of μ. An approximate 95% central credible interval for μ is defined by the 2.5% and 97.5% percentiles of the empirical distribution of the $\mu^{[m]}$. This approximation can be made as accurate as required by setting M sufficiently large.

3.1.2 Model comparisons

To test the null hypothesis $H_1 : \mu = \mu_1$ against the alternative $H_2 : \mu$ unspecified, we compute the likelihood ratio and deviance difference:

$$\begin{aligned}
LR_{12} &= \frac{c \cdot \frac{1}{\sigma^n} \exp\left\{-\frac{1}{2\sigma^2}[RSS + n(\bar{y} - \mu_1)^2]\right\}}{c \cdot \frac{1}{\sigma^n} \exp\left\{-\frac{1}{2\sigma^2}[RSS + n(\bar{y} - \mu)^2]\right\}} \\
&= \exp\left\{-\frac{1}{2\sigma^2}[n(\bar{y} - \mu_1)^2 - n(\bar{y} - \mu)^2]\right\}
\end{aligned}$$

$$D_{12} = \frac{1}{\sigma^2}[n(\bar{y} - \mu_1)^2 - n(\bar{y} - \mu)^2]$$

$$= \frac{n(\bar{y} - \mu_1)^2}{s^2} \cdot \frac{s^2}{\sigma^2} - \frac{n(\bar{y} - \mu)^2}{\sigma^2}$$

$$= t^2 \cdot W/(n-1) - Z^2$$

where t is the frequentist one-sample t-statistic, $W = RSS/\sigma^2$ has a χ^2_{n-1} distribution, and Z has an independent standard normal distribution.

Thus the posterior distribution of D_{12} is that of the difference between a $\chi^2_{n-1}/(n-1)$ variate scaled by t^2, and an independent χ^2_1 variate. This distribution has no closed form density, but can be directly simulated by making M independent draws from the two χ^2 variates and forming the M draws from the scaled difference D_{12}.

3.1.3 Example

We modify slightly an example from Gönen et al. (2005). In a sample of $n = 10$, we observe a sample mean $\bar{y} = 5$ with a sample standard deviation $s = 8.74$. The null hypothesis is $\mu = 0$, the alternative is $\mu \neq 0$. The t-statistic is 1.809, with a p-value of 0.104, well above the conventional criterion for rejection. The 95% credible interval for μ is $\bar{y} \pm 2.262s/\sqrt{n}$, here $(-1.252, 11.252)$. This is identical to the 95% confidence interval for μ (with the diffuse prior used).

We make 10,000 draws from the posterior distribution of D as described in Section 3.1.2. The cdf of these values is shown in Figure 3.1, and a smooth approximation to the density in Figure 3.2.

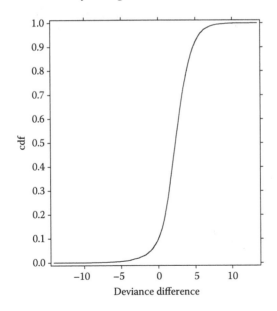

FIGURE 3.1
Posterior cdf of D_{12}.

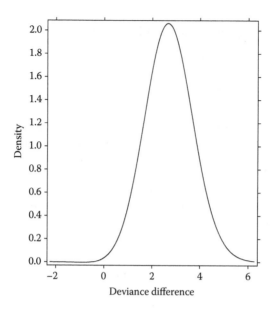

FIGURE 3.2
Posterior density of D_{12}.

The empirical probability that $D_{12} < 0$ (that is, that the null hypothesis has higher likelihood than the alternative) is 0.103, with simulation standard error 0.003. This is very close to the p-value, as expected from the Dempster fundamental relation. The empirical 95% central credible interval for D_{12} is $(-2.258, 6.266)$. The equivalent interval for the likelihood ratio is $(0.044, 3.09)$ and that for the posterior probability of the null hypothesis (with equal prior probabilities) is $(0.042, 0.756)$. The sample data are consistent with a wide range of values for μ, and hence for the likelihood ratio.

3.2 Two samples: equal variances

3.2.1 Credible interval

The two samples of sizes n_j, where $j = 1, 2$, provide sample means $\bar{y}_j$ and the pooled within-sample sum of squares

$$WSS = \sum_{i=1}^{n_1}(y_i - \bar{y}_1)^2 + \sum_{i=n_1+1}^{n_1+n_2}(y_i - \bar{y}_2)^2.$$

The diffuse prior distribution for μ_1, μ_2, and σ is $1/\sigma^2$. As for the one-sample case, the joint posterior distribution of (μ_1, μ_2, σ) can be factored in the form

$$\mu_j \mid \sigma, \mathbf{y} \sim \text{independent } N(\bar{y}_j, \sigma^2/n_j), \; WSS/\sigma^2 \mid \mathbf{y} \sim \chi^2_{n-2},$$

so that

$$\mu_1 - \mu_2 \mid \sigma, \mathbf{y} \sim N(\bar{y}_1 - \bar{y}_2, \sigma^2(1/n_1 + 1/n_2)),$$

and

$$[\mu_1 - \mu_2 - (\bar{y}_1 - \bar{y}_2)]/[\sigma\sqrt{1/n_1 + 1/n_2}] \mid \sigma, \mathbf{y} \sim N(0, 1).$$

Since this distribution does not depend on σ, the distribution holds unconditionally as well, and hence

$$t = [\mu_1 - \mu_2 - (\bar{y}_1 - \bar{y}_2)]/[s\sqrt{1/n_1 + 1/n_2}] \mid \mathbf{y} \sim t_{n-2}$$

where $n = n_1 + n_2$ and $s^2 = WSS/(n-2)$. The $100(1-\alpha)\%$ credible interval for $\mu_1 - \mu_2$ corresponds exactly, for the diffuse prior, to the usual $100(1-\alpha)\%$ t frequentist confidence interval.

As in the one-sample case, simulation methods can be used to generate a random sample from the posterior distribution of $\mu_1 - \mu_2$. We do not give details.

3.2.2 Model comparisons

The null hypothesis is $H_1 : \mu_1 = \mu_2$, the alternative is $H_2 : \mu_1 \neq \mu_2$. A Bayesian problem occurs here: we need to specify the form of the *nuisance parameter* corresponding to the parameter of interest, since we have to express the likelihood ratio in terms of the parameters defined under the alternative model: under the null hypothesis the common mean has to be expressed in terms of μ_1 and μ_2. In the frequentist maximized likelihood ratio approach this is not necessary, since maximization carries over to any transformations of the parameters.

We choose wherever possible the *orthogonal parametrization* (in the information matrix). It is easily verified that if we define $\theta = \mu_1 - \mu_2$ and $\phi = (n_1\mu_1 + n_2\mu_2)/(n_1 + n_2)$, the information matrix in θ, ϕ and σ is diagonal, so in large samples the joint posterior distribution of these parameters will have near joint independence. So we define $\mu = (n_1\mu_1 + n_2\mu_2)/n$ as the common mean under the null hypothesis.

We note for later reference that ϕ is the population form of the *MLE of the common mean* $\bar{y} = (n_1\bar{y}_1 + n_2\bar{y}_2)/n$ *under the null hypothesis*. This relationship occurs in many examples.

The likelihood under the alternative hypothesis is

$$L(\mu_1, \mu_2, \sigma) = \frac{1}{[\sqrt{2\pi}\sigma]^n} \exp\left\{-\frac{1}{2\sigma^2}[WSS + n_1(\bar{y}_1 - \mu_1)^2 + n_2(\bar{y}_2 - \mu_2)^2]\right\}$$

and the likelihood ratio and deviance difference are

$$LR_{12} = \exp\left\{-\frac{1}{2\sigma^2}[n_1(\bar{y}_1 - \mu)^2 - n_1(\bar{y}_1 - \mu_1)^2 + n_2(\bar{y}_2 - \mu)^2 - n_2(\bar{y}_2 - \mu_2)^2]\right\}$$

$$D_{12} = \frac{1}{\sigma^2}[n_1(\bar{y}_1 - \mu_1)^2 - n_1(\bar{y}_1 - \mu)^2 + n_2(\bar{y}_2 - \mu_2)^2 - n_2(\bar{y}_2 - \mu)^2]$$

$$= \frac{1}{\sigma^2}\frac{n_1 n_2}{n}\left\{(\bar{y}_1 - \bar{y}_2)^2 - [(\mu_1 - \mu_2) - (\bar{y}_1 - \bar{y}_2)]^2\right\}$$

$$= \frac{n_1 n_2}{n}\frac{(\bar{y}_1 - \bar{y}_2)^2}{s^2} \cdot \frac{s^2}{\sigma^2} - Z^2$$

$$= t^2 W/(n-2) - Z^2$$

where t is the two-sample t-statistic, $W = WSS/\sigma^2$ and has a χ^2_{n-2}, distribution, and

$$Z = \sqrt{\frac{n_1 n_2}{n}}\frac{(\mu_1 - \mu_2) - (\bar{y}_1 - \bar{y}_2)}{\sigma}$$

and has a standard normal distribution independent of W. These results are exactly parallel to those for the one-sample case.

3.2.3   Example

We use the example in Gönen et al. (2005), which we also adapted for the one-sample case above. In samples of $n_1 = 10$ and $n_2 = 11$, we observe sample means $\bar{y}_1 = 5.0$ and $\bar{y}_2 = -0.2727$, with sample standard deviations of $s_1 = 8.74$ and $s_2 = 5.90$. The two-sample t-statistic is 1.634, with a p-value of 0.119. The 95% credible interval for $\mu_1 - \mu_2$ is $\bar{y}_1 - \bar{y}_2 \pm 2.262 s\sqrt{n/(n_1 n_2)}$, here $(-1.025, 12.479)$. This is identical to the 95% confidence interval (with the diffuse priors used).

We make 10,000 draws from the posterior distribution of D_{12} as described in Section 3.1.2. The cdf of these values is shown in Figure 3.3, and a smooth approximation to the density in Figure 3.4.

The empirical probability that $D_{12} < 0$ is 0.117, with simulation standard error 0.003. This is again very close to the p-value, for the same reason given above. The empirical 95% equal-tailed credible interval for D_{12} is $(-2.576, 4.148)$. This interval is less wide than for the one-sample t-test because the larger sample size gives a better definition of σ. The equivalent interval for the likelihood ratio is $(0.126, 3.626)$, and that for the posterior probability of the null hypothesis is $(0.112, 0.784)$.

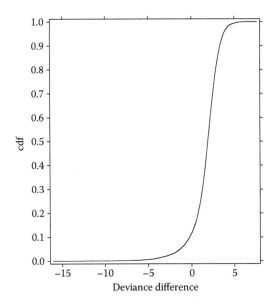

FIGURE 3.3
Posterior cdf of D_{12}.

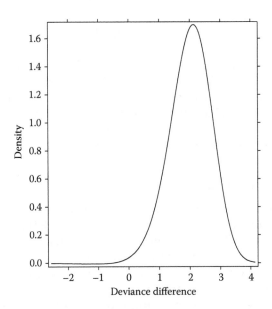

FIGURE 3.4
Posterior density of D_{12}.

3.3   A Bayes factor two-sample test

The Bayesian test proposed by Gönen et al. (2005) reparametrizes from (μ_1, μ_2, σ) to (μ, δ, σ), where $\mu = (\mu_1 + \mu_2)/2$ and δ is the effect size: $\delta = (\mu_2 - \mu_1)/\sigma$. They use noninformative priors for the common mean and σ: $\pi(\mu, \sigma) = c/\sigma^2$. The (independent) prior for the effect size is normal, with mean and variance hyper parameters, λ and σ_δ^2. These have to be set by the user of the test.

For the computation of the Bayes factor, the likelihood under each hypothesis is integrated over μ and σ; the difficulty that this prior is improper is finessed by the device of using a proper prior which is made to approach the limiting improper prior, and requiring this limiting process to be the same under each hypothesis, so that the integrating constant in the proper prior cancels out.

The form of the Bayes factor (BF) resulting is (Gönen et al., p. 253)

$$BF = \frac{T_v(t|0, 1)}{T_v(t|n_\delta^{1/2}\lambda, 1 + n_\delta\sigma_\delta^2)},$$

where t is the value of the frequentist two-sample test statistic, $n_\delta^{-1} = n_1^{-1} + n_2^{-1}$, and $T_v(.|a, b)$ denotes the noncentral t probability density function having location a, scale $b^{1/2}$, and degrees of freedom v. The integrated likelihood under the alternative and the Bayes factor depend explicitly on the values of the hyper parameters.

For the hyper parameter values $\lambda = 0$ and $\sigma_\delta = 1/3$, the Bayes factor is 0.791, equivalent to a posterior probability of H_0 of 0.442. The posterior probability is sensitive to both λ and σ_δ: Figures 1 and 2 in Gönen et al. (2005) show that this probability varies from 0.25 to 1.0 over the ranges of parameter values $\lambda \in (-1, 1)$, $\sigma_\delta \in 0.01, 1.00)$ considered. For comparison, the 95% credible interval for this posterior probability from Section 3.2.1 is (0.112, 0.784).

3.3.1   Informative prior for the effect size

The choice of the effect size as the parameter of interest, and its informative prior, greatly complicate the analysis because the three new model parameters are no longer information-orthogonal. The hyper parameters in the effect size prior have to be specified by the user; it is certainly not clear from its formulation how informative the prior is relative to the information in the likelihood. We can evaluate the latter fairly easily.

From the standard analysis for the two-sample problem earlier we have that for diffuse priors on μ_1 and μ_2, the posterior distribution of $\mu_2 - \mu_1$ given σ is $N(\bar{y}_2 - \bar{y}_1, \sigma^2(1/n_1 + 1/n_2))$, and therefore the posterior distribution of δ

given σ is

$$\delta \mid \sigma \sim N\left(\frac{\bar{y}_2 - \bar{y}_1}{\sigma}, \frac{1}{n_1} + \frac{1}{n_2}\right)$$
$$\sim N\left(\frac{\bar{y}_2 - \bar{y}_1}{s} \cdot \frac{s}{\sigma}, \frac{1}{n_1} + \frac{1}{n_2}\right)$$

where $s/\sigma \sim \sqrt{\chi_\nu^2/\nu}$, so that with increasing ν,

$$E\left[\frac{s}{\sigma}\right] \rightarrow 1,$$
$$\text{Var}\left[\frac{s}{\sigma}\right] \rightarrow 0.$$

The conditional (given σ) variance of δ in the example is $1/10 + 1/11 = 0.10909$; the conditional posterior standard deviation is 0.330. The unconditional variance is increased slightly by the variance of the conditional mean.

The range of prior standard deviations for δ considered by Gönen et al. (2005) is (0.01, 1.00). For one-third of this range the prior for δ is more informative than the likelihood; for the example they quote, of $\lambda = 0$, $\sigma_\delta = 1/3$, the prior will be more informative than the likelihood so the posterior mean for δ will be shrunk toward zero. This is reflected in the small Bayes factor of 0.791.

We are not arguing against the formation by subject-matter experts of informative priors for the effect size. But it is important for the expert, and even more for the novice, to know how the information in the prior and the likelihood are combined in the final analysis. For this reason we use wherever possible *diffuse* priors, to allow the information *in the data* to be assessed. Our approach does not require hyper parameters to be assigned or assessed, nor does it require the use of effect sizes.

3.4 Two samples: different variances

3.4.1 Credible interval

The two samples of sizes n_j, where $j = 1, 2$, now provide sample means $\bar{y}_j$, and sums of squares $RSS_j = \sum_{i=1}^{n_j}(y_i - \bar{y}_j)^2$. The noninformative prior distribution for μ_1, μ_2, σ_1, and σ_2 is $1/(\sigma_1\sigma_2)$. As for the one-sample case, the joint posterior distribution of $(\mu_1, \mu_2, \sigma_1, \sigma_2)$ can be factored in the form

$$\mu_j \mid \sigma_j, \mathbf{y} \sim N(\bar{y}_j, \sigma_j^2/n_j), \quad RSS_j/\sigma_j^2 \mid \mathbf{y} \sim \chi_{n_j-1}^2,$$

so that

$$\mu_1 - \mu_2 \mid \sigma_1, \sigma_2, \mathbf{y} \sim N(\bar{y}_1 - \bar{y}_2, \sigma_1^2/n_1 + \sigma_2^2/n_2).$$

Integrating out σ_1 and σ_2 to obtain the exact analytic posterior is complicated (the Behrens-Fisher distribution for the difference between scaled t variables), but simulation methods can be used very easily to generate a sample from the posterior distribution of $\mu_1 - \mu_2$:

- Make M random draws of RSS_j/σ_j^2, where $j = 1, 2$ from their marginal $\chi_{n_j-1}^2$ distributions, and hence M random draws $\sigma_j^{[m]}$ of σ_j.
- For each $\sigma_j^{[m]}$, make a random draw $\mu_j^{[m]}$ of μ_j from its conditional normal $N(\bar{y}_j, \sigma_j^{[m]2}/n_j)$ distribution.
- Compute $\delta^{[m]} = \mu_1^{[m]} - \mu_2^{[m]}$.

Alternatively (Tanner, 1996, p. 29) we can make M random draws from t_1 and t_2 and convert them to μ_1 and μ_2, and then δ.

Credible intervals for δ can be computed as above from the empirical distribution of the $\delta^{[m]}$. The credible intervals in this case correspond exactly, for the noninformative priors, to the fiducial Behrens-Fisher intervals (Tanner, 1996, pp. 20–21).

3.4.2   Model comparison

The null hypothesis is $H_1 : \mu_1 = \mu_2$, the alternative is $H_2 : \mu_1 \neq \mu_2$. As in the previous case of equal variances, we need to define the common mean under the null hypothesis. It is easily verified that the information orthogonalizing choice for the nuisance parameter is

$$\mu = \left[\frac{n_1\mu_1}{\sigma_1^2} + \frac{n_2\mu_2}{\sigma_2^2} \right] \bigg/ \left[\frac{n_1}{\sigma_1^2} + \frac{n_2}{\sigma_2^2} \right];$$

this is again the population form of the MLE of the common mean under the null hypothesis. The likelihood under the alternative hypothesis is

$$\frac{1}{[\sqrt{2\pi}]^n \sigma_1^{n_1} \sigma_2^{n_2}} \exp\left\{ -\frac{1}{2\sigma_1^2}[RSS_1 + n_1(\bar{y}_1 - \mu_1)^2] - \frac{1}{2\sigma_2^2}[RSS_2 + n_2(\bar{y}_2 - \mu_2)^2] \right\}$$

and the likelihood ratio and deviance difference are

$$LR_{12} = \exp\left\{ -\frac{n_1}{2\sigma_1^2}[(\bar{y}_1 - \mu)^2 - (\bar{y}_1 - \mu_1)^2] - \frac{n_2}{2\sigma_2^2}[(\bar{y}_2 - \mu)^2 - (\bar{y}_2 - \mu_2)^2] \right\}$$

$$D_{12} = \frac{n_1}{\sigma_1^2}[(\bar{y}_1 - \mu)^2 - (\bar{y}_1 - \mu_1)^2] + \frac{n_2}{\sigma_2^2}[(\bar{y}_2 - \mu)^2 - (\bar{y}_2 - \mu_2)^2]$$

$$= \left\{ (\bar{y}_1 - \bar{y}_2)^2 - [(\mu_1 - \mu_2) - (\bar{y}_1 - \bar{y}_2)]^2 \right\} / \left[\frac{\sigma_1^2}{n_1} + \frac{\sigma_2^2}{n_2} \right]$$

$$= \frac{(\bar{y}_1 - \bar{y}_2)^2}{\frac{s_1^2}{n_1} + \frac{s_2^2}{n_2}} \cdot \frac{\frac{s_1^2}{n_1} + \frac{s_2^2}{n_2}}{\frac{\sigma_1^2}{n_1} + \frac{\sigma_2^2}{n_2}} - Z^2$$

$$= t^{*2} \cdot \frac{\frac{s_1^2}{n_1} + \frac{s_2^2}{n_2}}{\frac{\sigma_1^2}{n_1} + \frac{\sigma_2^2}{n_2}} - Z^2$$

where

$$Z = \frac{(\mu_1 - \mu_2) - (\bar{y}_1 - \bar{y}_2)}{\sqrt{\frac{\sigma_1^2}{n_1} + \frac{\sigma_2^2}{n_2}}}$$

has a standard normal distribution independent of σ_1^2 and σ_2^2,

$$t^* = \frac{\bar{y}_1 - \bar{y}_2}{\sqrt{\frac{s_1^2}{n_1} + \frac{s_2^2}{n_2}}}$$

is the Welch t^* statistic proposed as an approximate t-test alternative to the Behrens-Fisher test, and $s_j^2 = RSS_j/(n_j - 1)$. The distribution of D_{12} has no closed form density, but can be directly simulated by generating M independent values of the two χ^2 variates, RSS_1/σ_1^2 and RSS_2/σ_2^2, and of Z, and forming the M values of D_{12}.

3.4.3 Example

We use the same example but now do not assume the variances are equal. We compare the Bayes analysis with the Welch approximation which uses t^* with fractional degrees of freedom ν:

$$\nu = \left[\frac{s_1^2}{n_1} + \frac{s_2^2}{n_2} \right]^2 / \left[\frac{(s_1^2/n_1)^2}{n_1 - 1} + \frac{(s_2^2/n_2)^2}{n_2 - 1} \right].$$

The t^* statistic is 1.742, with approximate degrees of freedom 15.59, and approximate p-value of 0.102.

We make 10,000 random draws from the posterior distribution of D_{12} as described in Section 3.1.2. The cdf of these values is shown in Figure 3.5, and a smooth approximation to the density in Figure 3.6.

The empirical probability that $D_{12} < 0$ is 0.144, with simulation standard error 0.004. The empirical 95% equal-tailed credible interval for D_{12} is $(-2.782, 3.842)$, and the intervals for the likelihood ratio and posterior probability of the null hypothesis are $(0.147, 4.02)$ and $(0.128, 0.801)$. These intervals give even

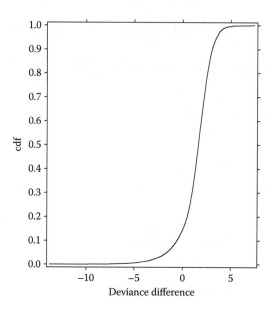

FIGURE 3.5
Posterior cdf of D_{12}.

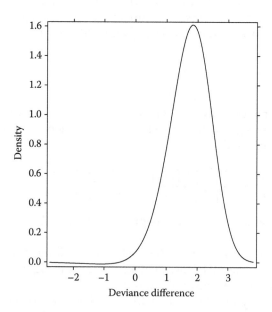

FIGURE 3.6
Posterior density of D_{12}.

less evidence against the null hypothesis, as expected from the weaker model assumption of different variances. The empirical probability that $LR < 1/9$ (i.e., that $D_{12} > 2\log 9 = 4.394$) is 0.009.

The Bayesian version gives somewhat more support to the null hypothesis than the Welch test.

3.5 The normal model variance

3.5.1 Credible interval

We have a random sample $\mathbf{y} = (y_1, \ldots, y_n)$ from the normal model $Y \sim N(\mu, \sigma^2)$, and want a credible interval for the variance. With the diffuse priors for μ and σ, the joint posterior distribution is given in Section 3.1.1:

$$\mu \mid \sigma, \mathbf{y} \sim N(\bar{y}, \sigma^2/n), \quad RSS/\sigma^2 \mid \mathbf{y} \sim \chi_{n-1}^2.$$

Integrating out μ, the marginal posterior distribution of σ^2 is given by $RSS/\sigma^2 \sim \chi_{n-1}^2$. The component of the likelihood involving RSS is called in frequentist theory the *marginal* or *restricted likelihood* for σ^2, though in this theory its use can be justified formally only by an appeal to the use of the $n-1$ independent *linear contrasts* of the observations whose distributions do not depend on μ, or by the Bayesian argument of integrating out μ from the likelihood with a flat prior.

The $100(1-\alpha)\%$ central credible interval for σ^2 with these priors is

$$[RSS/\chi_{n-1,1-\alpha/2}^2 < \sigma^2 < RSS/\chi_{n-1,\alpha/2}^2],$$

which is identical to the $100(1-\alpha)\%$ equal-tailed confidence interval for σ^2 based on the marginal likelihood.

3.5.2 Model comparisons

We wish to test a null hypothesis, $H_1 : \sigma^2 = \sigma_1^2$, against a general alternative, in the normal model $Y \sim N(\mu, \sigma^2)$, given data $\mathbf{y} = (y_1, \ldots, y_n)$.

In frequentist testing, we use the likelihood ratio test. The ratio of maximized likelihoods and the likelihood ratio test statistic are easily seen to be:

$$MLR_{12} = \frac{L_1(\bar{y}, \sigma_1^2)}{L_2(\bar{y}, \hat{\sigma}^2)} = \left[\frac{\hat{\sigma}^2}{\sigma_1^2}\right]^{n/2} \exp\left\{-\frac{1}{2}\left[\frac{RSS}{\sigma_1^2} - n\right]\right\}$$

$$LRTS = -2\log MLR_{12} = -n\left[\log\frac{\hat{\sigma}^2}{\sigma_1^2} - \frac{\hat{\sigma}^2}{\sigma_1^2} + 1\right],$$

where $\hat{\sigma}^2 = RSS/n$. The asymptotic distribution of the likelihood ratio test statistic (LRTS) under the null hypothesis is χ_1^2.

However, historically this test has been little used, because it does not easily provide a confidence interval for σ^2, as it is not a monotone function of σ^2. This is provided by using directly the exact (sampling) distribution of RSS/σ^2. Large *or* small values of $\hat{\sigma}^2/\sigma_1^2$ will give large values of the LRTS, so the frequentist LR test is equivalent to using the ratio RSS/σ_1^2 as the test statistic, and rejecting the null hypothesis with a size α test if this ratio exceeds the $100(1-\alpha/2)$ percentile, or is less than the $100\alpha/2$ percentile of χ_{n-1}^2. This is equivalent to rejecting if the null value σ_1^2 falls outside the equal-tailed confidence (or credible) interval for σ^2 in Section 3.6.1.

If we use the marginal likelihood for the corresponding Bayesian test, we have the marginal null and alternative likelihoods (omitting numerical constants which cancel), and their ratio and deviance difference:

$$M_1 = M(\sigma_1) = \frac{1}{\sigma_1^\nu} \exp\left\{-\frac{RSS}{2\sigma_1^2}\right\}$$

$$M_2 = M(\sigma) = \frac{1}{\sigma^\nu} \exp\left\{-\frac{RSS}{2\sigma^2}\right\}$$

$$ML\,R_{12} = M_1/M_2 = \left[\frac{\sigma^2}{\sigma_1^2}\right]^{\nu/2} \exp\left\{-\frac{RSS}{2}\left[\frac{1}{\sigma_1^2} - \frac{1}{\sigma^2}\right]\right\}$$

$$MD_{12} = -2\log ML\,R_{12} = -\nu\log\left[\frac{\sigma^2}{\sigma_1^2}\right] + \frac{RSS}{\sigma_1^2} - \frac{RSS}{\sigma^2}$$

$$= -\nu\log\left[\frac{RSS/\sigma_1^2}{RSS/\sigma^2}\right] + \frac{RSS}{\sigma_1^2} - \frac{RSS}{\sigma^2}$$

$$= -\nu(\log w_1 - \log W) + w_1 - W,$$

where $w_1 = RSS/\sigma_1^2$, $W = RSS/\sigma^2$ has a posterior χ_ν^2 distribution, and $\nu = n-1$. The deviance difference has no closed-form distribution, but is easily simulated.

3.5.3 Example

In a sample of 20, we have the sample standard deviation $s = 4$ ($RSS = 304$), with a hypothesis $H_1 : \sigma = 3$. The value of $w_1 = RSS/\sigma_1^2$ is 33.8, which exceeds the 97.5th percentile (32.85) of χ_{19}^2; the 95% central credible interval for σ^2 is $(9.25, 34.13)$ which excludes the null value 9.

This approach leads to rejection of the null hypothesis by a 5% size test. If the LRTS is used instead, the value of the test statistic is

$$-20(\log(15.2/9) - 15.2/9 + 1) = 3.29.$$

This does *not* exceed the 95th percentile (3.84) of χ_1^2; its *p*-value is 0.070. So this test would *not* reject the null hypothesis. The reason for this discrepancy

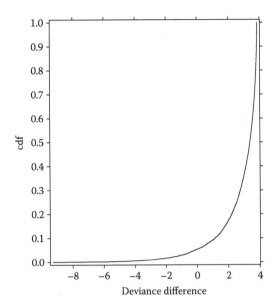

FIGURE 3.7
Posterior distribution of marginal deviance difference.

is that the likelihood in the parameter σ^2 is not symmetric, so it is not clear that an equal-tailed rejection region is appropriate. The LRTS implies an unequal choice of tail probabilities for the statistic RSS/σ^2.

Our interest, however, is in the Bayesian test; for the test based on the marginal likelihood, we have $-\nu \log w_1 + w_1 = -33.09$. We make $M = 10,000$ draws $W^{[m]}$ from $W \sim \chi^2_{19}$, and compute the M values $MD_{12}^{[m]} = -33.09 + \nu \log W^{[m]} - W^{[m]}$. The empirical cdf of MD_{12} is shown in Figure 3.7. (Here the marginal posterior deviance is closely approximated by its asymptotic distribution: $3.855 - \chi^2_1$), though this is not the full deviance).

The empirical probability that $MD_{12} > 0$ is 0.053 with SE 0.0022. The evidence against the null hypothesis assessed directly by the marginal likelihood ratio is slightly weaker than that from the credible interval, but stronger than from the LRTS.

3.5.4 Marginal and full likelihoods

Why do we use the restricted likelihood for this problem? We have the *full* likelihood, which we use in all other model applications: why is this problem different? The additional term in the full likelihood appears to be able to contribute additional information about σ, though it has not been clear how this could be done.

The full likelihoods under the null and alternative hypotheses, and their ratio and the deviance difference are, in the same notation:

$$L_1 = L(\mu, \sigma_1) = \left[\frac{1}{\sigma_1}\right]^n \exp\left\{-\frac{1}{2}\left[\frac{RSS + n(\bar{y} - \mu)^2}{\sigma_1^2}\right]\right\}$$

$$L_2 = L(\mu, \sigma) = \left[\frac{1}{\sigma}\right]^n \exp\left\{-\frac{1}{2}\left[\frac{RSS + n(\bar{y} - \mu)^2}{\sigma^2}\right]\right\}$$

$$LR_{12} = L_1/L_2 = \left[\frac{\sigma^2}{\sigma_1^2}\right]^{n/2}$$

$$\times \exp\left\{-\frac{1}{2}\left[\frac{RSS}{\sigma_1^2} + \frac{n(\bar{y} - \mu)^2}{\sigma_1^2} - \frac{RSS}{\sigma^2} - \frac{n(\bar{y} - \mu)^2}{\sigma^2}\right]\right\}$$

$$D_{12} = -2\log LR_{12} = -n\log\left[\frac{RSS/\sigma_1^2}{RSS/\sigma^2}\right] + \frac{RSS}{\sigma_1^2} + \frac{n(\bar{y} - \mu)^2}{\sigma^2}\left[\frac{RSS/\sigma_1^2}{RSS/\sigma^2}\right]$$

$$-\frac{RSS}{\sigma^2} - \frac{n(\bar{y} - \mu)^2}{\sigma^2}$$

$$= -n(\log w_1 - \log W) + w_1 - W + Z^2(w_1/W - 1),$$

where $Z^2 = n(\bar{y} - \mu)^2/\sigma^2$.

This differs from MD_{12} by the additional degree of freedom in the n multiplier, and the term in Z^2; the difference is

$$D_{12} - MD_{12} = \log(w_1/W) + Z^2(w_1/W - 1).$$

So when the W draws exceed w_1, the difference will be negative, and when the draws are less than w_1, the difference will be positive. Figure 3.8 shows the two deviance distributions for the same example with $\sigma_1 = 3$ and sample standard deviation $s = 4$, in a sample of $n = 20$ using $M = 10,000$ draws.

The full-deviance cdf is solid, the marginal-deviance cdf is dotted. They behave quite differently in the upper tail: the marginal deviance difference has a maximum over all draws when $W = v$ which in this example is 3.855.

There is no corresponding upper limit on the full deviance, because of the additional term: Z^2 can be quite large – up to 10 in 10,000 samples – and if the W draw is small when the Z draw is large (positive or negative), the second term can be substantial, greatly increasing the tail of the full deviance distribution. (This distribution cannot be represented by its asymptotic form, which is $3.29 - \chi_1^2$. In large samples the effect of the term in Z^2 is dominated by the n term, but in small samples its effect can be large.)

The empirical probability that the deviance difference from the full likelihood is positive is 0.0508, compared with 0.0533 from the marginal likelihood. The difference is small, but real. The median deviances are very similar: 3.35 for the full, 3.39 for the marginal, but the 95% central credible intervals are quite different, as expected from the long tail of the full deviance: $[-1.20, 7.66]$ for the full, but $[-1.23, 3.84]$ for the marginal.

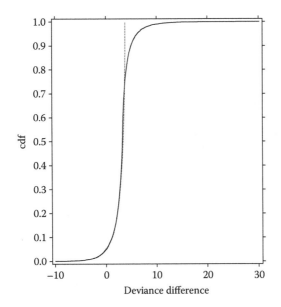

FIGURE 3.8
Posterior distributions of D_{12} and MD_{12}: variance test.

So the full likelihood *provides stronger evidence against the null hypothesis* than the marginal likelihood, in the sense of giving posterior weight to larger values of the deviance difference than can the marginal likelihood.

Using the likelihood distribution resolves the argument over whether the full or the marginal likelihood should be used. From the posterior likelihood distribution point of view, one should always use the full likelihood (as in all other cases!), since the term neglected in the marginal likelihood provides additional evidence which can change the inference about the variance parameter.

3.6   Variance heterogeneity test

We extend the previous analysis to the two-sample case. We note first that the standard frequentist *variance ratio test* for variance homogeneity is very straightforward: we compute the variance ratio

$$F = \frac{RSS_1/(n_1 - 1)}{RSS_2/(n_2 - 1)} = s_1^2/s_2^2$$

which has an F_{n_1-1,n_2-1} repeated sampling distribution under the null hypothesis of a common variance. Large or small values of the observed F lead to rejection; it is conventional to order the sample variances so that the observed ratio is greater than 1; the ratio is then compared with the $100(1 - \alpha/2)$ percentile of the appropriate F distribution.

We note, however, that this test is not the likelihood ratio test; the LRTS is easily shown to be

$$LRTS = -2\log(L_{1max}/L_{2max}) = n\log\hat{\sigma}^2 - (n_1\log\hat{\sigma}_1^2 + n_2\log\hat{\sigma}_2^2).$$

Its use may give a different conclusion as it again implies different tail-area probabilities for the ratio of variances.

For the Bayesian test, as before, the two samples of sizes n_j provide sample means $\bar{y}_j$ and sums of squares $RSS_j = \sum_{i=1}^{n_1}(y_i - \bar{y}_j)^2$, $j = 1, 2$. The noninformative prior distribution for μ_1, μ_2, σ_1, and σ_2 is $1/(\sigma_1\sigma_2)$. The null hypothesis is $H_1 : \sigma_1 = \sigma_2$, the alternative is $H_2 : \sigma_1 \neq \sigma_2$. The likelihoods and deviances under the null and alternative are

$$L_1 = L(\mu_1, \mu_2, \sigma)$$
$$= c \cdot \frac{1}{\sigma^n}\exp\left\{-\frac{1}{2\sigma^2}\left[n_1(\bar{y}_1 - \mu_1)^2 + n_2(\bar{y}_2 - \mu_2)^2 + RSS_1 + RSS_2\right]\right\}$$
$$L_2 = L(\mu_1, \mu_2, \sigma_1, \sigma_2)$$
$$= c \cdot \frac{1}{\sigma_1^{n_1}}\exp\left\{-\frac{1}{2\sigma_1^2}\left[n_1(\bar{y}_1 - \mu_1)^2 + RSS_1\right]\right\}$$
$$\cdot \frac{1}{\sigma_2^{n_2}}\exp\left\{-\frac{1}{2\sigma_2^2}\left[n_2(\bar{y}_2 - \mu_2)^2 + RSS_2\right]\right\}$$
$$D_1 = n\log\sigma^2 + \frac{1}{\sigma^2}\left[n_1(\bar{y}_1 - \mu_1)^2 + n_2(\bar{y}_2 - \mu_2)^2 + RSS_1 + RSS_2\right]$$
$$D_2 = n_1\log\sigma_1^2 + n_2\log\sigma_2^2 + \frac{1}{\sigma_1^2}\left[n_1(\bar{y}_1 - \mu_1)^2 + RSS_1\right]$$
$$+ \frac{1}{\sigma_2^2}\left[n_2(\bar{y}_2 - \mu_2)^2 + RSS_2\right].$$

We define the common variance under H_1 by the population form of the MLE:

$$\sigma^2 = (n_1\sigma_1^2 + n_2\sigma_2^2)/n,$$

with $n = n_1 + n_2$.

Using the same diffuse priors as in Section 3.3.1, we have the same posteriors under the alternative hypothesis:

$$\mu_j \mid \sigma_j, \mathbf{y} \sim N(\bar{y}_j, \sigma_j^2/n_j), \ RSS_j/\sigma_j^2 \mid \mathbf{y} \sim \chi_{n_j-1}^2,$$

so that

$$Z_j^2 = \frac{n_j(\bar{y}_j - \mu_j)^2}{\sigma_j^2} \sim \chi_1^2$$

independently of $W_j = RSS_j/\sigma_j^2 \sim \chi^2_{n_j-1}$. The distribution of the deviance difference then follows from

$$D_2 = n_1 \log \sigma_1^2 + n_2 \log \sigma_2^2 + W_1 + Z_1^2 + W_2 + Z_2^2$$

$$D_1 = n \log \sigma^2 + \frac{\sigma_1^2}{\sigma^2} \left[W_1 + Z_1^2 \right] + \frac{\sigma_2^2}{\sigma^2} \left[W_2 + Z_2^2 \right]$$

$$D_{12} = n \log \sigma^2 - [n_1 \log \sigma_1^2 + n_2 \log \sigma_2^2] + \left(\frac{\sigma_1^2}{\sigma^2} - 1 \right) \left[W_1 + Z_1^2 \right]$$

$$+ \left(\frac{\sigma_2^2}{\sigma^2} - 1 \right) \left[W_2 + Z_2^2 \right].$$

We make M independent draws $W_1^{[m]}, W_2^{[m]}, Z_1^{[m]}, Z_2^{[m]}$ from, respectively, $\chi^2_{n_1-1}$, $\chi^2_{n_2-1}$, $N(0, 1)$, and $N(0, 1)$, form the corresponding draws $\sigma_1^{[m]2} = RSS_1/W_1^{[m]}$, $\sigma_2^{[m]2} = RSS_2/W_2^{[m]}$ and $\sigma^{[m]2} = (n_1\sigma_1^{[m]2} + n_2\sigma_2^{[m]2})/n$, and substitute these in D_{12} to give the M draws

$$D_{12}^{[m]} = n \log \sigma^{[m]2} - [n_1 \log \sigma_1^{[m]2} + n_2 \log \sigma_2^{[m]2}] + \left(\frac{\sigma_1^{[m]2}}{\sigma^{[m]2}} - 1 \right) \left[W_1^{[m]} + Z_1^{[m]2} \right]$$

$$+ \left(\frac{\sigma_2^{[m]2}}{\sigma^{[m]2}} - 1 \right) \left[W_2^{[m]} + Z_2^{[m]2} \right].$$

For the two-group Gönen et al. (2005) example considered earlier, we have sample standard deviations $s_1 = 8.74$ and $s_2 = 5.90$ with sample sizes $n_1 = 10$, and $n_2 = 11$. The observed variance ratio is $F = 2.19$ with 9 and 10 degrees of freedom, which has an upper-tail probability of 0.119 (the two-tailed probability is 0.238). There is no evidence of variance heterogeneity. The likelihood ratio test gives a test statistic of 1.504 with an (asymptotic) p-value from χ_1^2 of 0.220, with the same conclusion.

For the Bayesian test, we show in Figure 3.9 the posterior cdf of the deviance difference based on $M = 10,000$ draws.

The empirical posterior probability that the deviance difference is negative is 0.199 with simulation SE 0.004, close to the LRTS p-value. The median deviance difference is 1.14, and the 95% central credible interval for the true deviance difference is $[-2.83, 4.36]$.

Figure 3.10 gives the empirical cdf of the log variance ratio $\log(\sigma_1^2/\sigma_2^2)$.

The 95% central credible interval for the variance ratio is $[0.58, 8.65]$. This includes 1, but is very wide: the small samples give little information about the variances and their ratio.

3.6.1 Nonrobustness of variance tests

It is well known in frequentist theory that tests on normal model variances are very sensitive to nonnormality of the population response. This applies

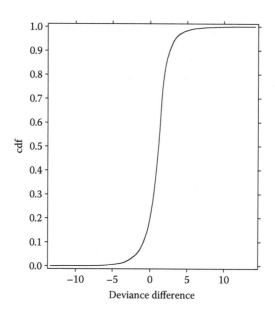

FIGURE 3.9
Posterior distribution of D_{12}: variance homogeneity test.

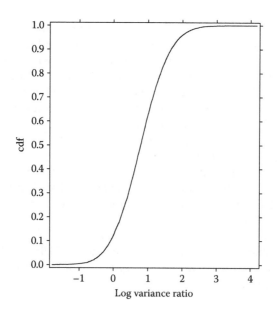

FIGURE 3.10
Posterior distribution of the log variance ratio.

equally to Bayesian tests since these use the same statistics and distributions as the frequentist theory. If we are particularly interested in inferences about variances, these can be carried out in a *distribution-free* or *nonparametric* way without relying on *any* assumption of a specific distribution of the response Y. This approach is described in detail in Chapter 4.

4

Unified Analysis of Finite Populations

This chapter is in one sense a diversion from the main theme of the book – the new approach to the Bayesian comparison of models. However, it addresses another major issue in statistical theory – the division (really a *gulf*) between the "design-based" and "model-based" schools of inference. Many statisticians, both Bayesian and design-based, have addressed this issue – see the recent example by Little and Zheng (2007) and the paper discussion. A book-length technical and philosophical discussion can be found in Brewer (2002).

The design-based school, which is followed by a substantial proportion of all statistical analysts, is concentrated in government and major survey sampling organizations, and so has a major impact on the analysis and presentation of official surveys. The design-based school bases statistical inference on probability calculations across *hypothetical replications of the survey design*. No model is used for *analysis*, though models may *guide* the form of population and subpopulation estimates used in the analysis. The terms *model-assisted* or *model-guided* have become popular to describe this approach (see for example, Särndal et al. 1992).

Since no model is used, there is no likelihood on which to base inference procedures. This has been a major criticism of the approach by model-based statisticians, for reasons which will be clear from the analysis in this chapter. *Any* model for the response variable will lead to optimal analysis (in the likelihood or Bayesian sense), and so the failure to use a model must lead to suboptimal analysis, and possibly misleading results.

The design-based school responds that since models are *simplifications* of the true population structure for mathematical convenience, they are *by definition wrong*, and if seriously wrong, can lead to *seriously misleading conclusions*. Model checking, a widespread requirement in model-based theory (discussed in Chapter 7), is time-consuming and essentially fruitless, since we are unable to *confirm* the correctness of a model; we can only *fail to disconfirm it*. Survey organizations have neither the time nor the resources to carry out extensive model checking, so it is far better to use procedures which have known inferential properties which do not depend on the correctness of a model, and so are *always correct* (at least in large samples), though they may be suboptimal relative to a model-based procedure, *if* the model were *known* to be correct.

These arguments have continued for many years without much change in positions. In this chapter we examine the model assumption, and develop a *model-free* or *nonparametric* approach to finite population analysis which can nevertheless be expressed as an *always true model*. This leads to a *full likelihood* and *fully Bayesian* (and maximum likelihood) analyses *which are optimal* in the model-based sense.

This approach is not new – it was first developed in the 1960s – but has lain unused, mainly because of the computational requirements for large-scale simulation, which were unavailable at that time. These constraints were removed long ago, and this approach can be directly implemented with available software and quite modest computing time. This chapter gives a full discussion, illustrated with an example of a complex clustered and stratified survey design from a standard survey sampling text book.

We give first a summary of the design-based approach. A clear and detailed treatment can be found in Lohr (1999). We then apply the approach of Chapter 2 to successively more complex problems in finite population survey sampling, following Aitkin (2008). We greatly extend the simple one-sample problem of Chapter 1.

> We are thus entering a rather narrow area of statistical theory, but it is an area which has been intensively cultivated, and this on the grounds of its practical importance rather than of its mathematical attractiveness. (Kendall and Stuart 1966, p. 166)

4.1 Sample selection indicators

We aim to determine the information about the population mean of a variable Y^* of interest, in a finite population of size N. We denote the values of Y^* by $Y_1^*, Y_2^*, \ldots, Y_N^*$, and define the *population mean* μ by

$$\mu = \sum_{I=1}^{N} Y_I^* / N$$

and the *population variance* by

$$\sigma^2 = \sum_{I=1}^{N} (Y_I^* - \mu)^2 / N.$$

In the survey literature, the variance denominator is usually $N-1$, for reasons which will appear below.

We consider first the case where N is large, either absolutely, or at least relative to the sample size n. This is the situation for the majority of sampled populations in large-scale surveys.

We draw a simple random sample (SRS) of fixed predetermined size n, and obtain observed values $y_1, \ldots, y_n$, with sample mean $\bar{y}$ and sample variance

$$s^2 = \sum_i (y_i - \bar{y})^2 / (n - 1).$$

We define indicator variables $Z_1, Z_2, \ldots, Z_I, \ldots, Z_N$: let

$$Z_I = 1 \text{ if population member } I \text{ is selected}$$
$$= 0 \text{ if population member } I \text{ is not selected.}$$

Then

$$\bar{y} = \sum_{i=1}^{n} y_i / n = \sum_{I=1}^{N} Y_I^* Z_I / \sum_{I=1}^{N} Z_I = \sum_{I=1}^{N} Y_I^* Z_I / n.$$

Inference about the mean μ is based on the *repeated sampling properties of the random variable $\bar{y}$* as an estimator of μ. These are evaluated by considering the Y_I^* as *fixed constants*, and the Z_I as Bernoulli *random variables*, with

$$\Pr[Z_I = 1] = \frac{\text{no. of samples containing unit I}}{\text{no. of samples of size } n}$$

$$= \frac{\binom{N-1}{n-1}}{\binom{N}{n}} = \frac{n}{N} = \pi,$$

the *sampling fraction*, and so

$$\mathrm{E}[Z_I] = \mathrm{E}[Z_I^2] = \pi, \ \mathrm{Var}[Z_I] = \pi(1 - \pi) = \frac{n}{N}\left(1 - \frac{n}{N}\right).$$

Hence

$$\mathrm{E}[\bar{y}] = \frac{1}{n}\sum_{I=1}^{N} Y_I^* \mathrm{E}[Z_I] = \frac{1}{N}\sum_{I=1}^{N} Y_I^* = \mu.$$

So as a random variable, $\bar{y}$ is *unbiased* for μ. To establish the variance of $\bar{y}$, we need the joint distribution of pairs of the Z_I. These are not independent; if we know that unit I is included in the sample, this reduces the probabilities that the other units are included:

$$\Pr[Z_I = 1, Z_J = 1] = \Pr[Z_I = 1]\Pr[Z_J = 1 \mid Z_I = 1]$$

$$= \frac{n}{N} \cdot \frac{n-1}{N-1}$$

and so

$$\text{Cov}[Z_I, Z_J] = \frac{n(n-1)}{N(N-1)} - \left(\frac{n}{N}\right)^2$$

$$= -\frac{1}{N-1}\frac{n}{N}\left(1 - \frac{n}{N}\right)$$

$$= -\pi(1-\pi)/(N-1).$$

Finally,

$$\text{Var}[\bar{y}] = \sum_I Y_I^{*2}\text{Var}[Z_I]/n^2 + \sum_I\sum_{\neq J} Y_I^* Y_J^*\text{Cov}[Z_I, Z_J]/n^2$$

$$= \frac{1-n/N}{nN(N-1)}\left[(N-1)\sum_I Y_I^{*2} - \sum_I\sum_{\neq J} Y_I^* Y_J^*\right]$$

$$= \frac{1-n/N}{n(N-1)}\sum_I(Y_I^* - \mu)^2$$

$$= (1-n/N)\sigma^2/n$$

$$= (1-\pi)\sigma^2/n$$

if the population variance is defined by the $N-1$ denominator. Whether N or $N-1$ is used makes no practical difference in large surveys. The first term is a *finite population correction*; if the sample fraction is small the variance of $\bar{y}$ is effectively σ^2/n, but it tends to zero as the sample size approaches the population size, and the sample exhausts the population. Similarly,

$$E[s^2] = E\left[\sum_i y_i^2 - n\bar{y}^2\right]/(n-1)$$

$$= E\left[\sum_I Z_I Y_I^{*2} - n\bar{y}^2\right]/(n-1)$$

$$= \left\{n\sum_I Y_I^{*2}/N - n(\text{Var}[\bar{y}] + E[\bar{y}]^2)\right\}/(n-1)$$

$$= \left\{n\sum_I Y_I^{*2}/N - n([1-1/N]\sigma^2/n + \mu^2)\right\}/(n-1)$$

$$= (1-1/N)\sigma^2.$$

So s^2 is an almost unbiased estimator of σ^2, regardless of any distribution for Y^*, and under the Bernoulli model, $\bar{y}$ is the minimum variance linear unbiased estimate of μ.

TABLE 4.1

Family Income Data, in Units of 1000 Dollars

26	35	38	39	42	46	47	47	47	52
53	55	55	56	58	60	60	60	60	60
65	65	67	67	69	70	71	72	75	77
80	81	85	93	96	104	104	107	119	120

However, for confidence interval statements about μ, we have to rely on the central limit theorem, that as $n \to \infty$ (and $N \to \infty$), the sampling distribution of

$$z = \frac{\sqrt{n}(\bar{y} - \mu)}{\sigma} \to N(0, 1)$$

as does that of

$$t = \frac{\sqrt{n}(\bar{y} - \mu)}{s},$$

giving the usual large-sample confidence interval

$$\bar{y} - z_{1-\alpha/2}s/\sqrt{n} < \mu < \bar{y} + z_{1-\alpha/2}s/\sqrt{n}.$$

The accuracy of the confidence interval coverage depends on the sample size n – it may be quite inaccurate for small n. Without other information about the Y^* population, we cannot say more.

For the income example of Chapter 1, we have $\bar{y} = 67.1, s^2 = 500.87$, and the (approximate) 95% confidence interval for the population mean is [60.1, 74.0]. The data are reproduced in Table 4.1.

4.2 The Bayesian bootstrap

The Bayesian bootstrap approach to inference in finite population survey sampling was described by Hartley and Rao (1968), Ericson (1969), and Rubin (1981), who gave it this name. This approach does not use a parametric model assumption for the distribution of the numerically valued response variable Y^* in the finite population of size N, taking values $Y_1^*, \ldots, Y_N^*$. We re-express these in terms of the *distinct* values which Y^* can take, denoted by $Y_1 < \cdots < Y_J < \cdots < Y_D$, where Y_1 is the smallest and Y_D the largest population values. As these values are always measured with finite precision, denoted by δ, the possible values of Y^* form an equally spaced discrete grid of values Y_J with step-length δ, with counts N_J, and proportions $p_J = N_J/N$ at Y_J; these counts can be obtained formally by a tabulation by the distinct values.

4.2.1   Multinomial model

The key to the Bayesian bootstrap approach is the *multinomial model* for the population, with proportions p_J at the values Y_J. Population parameters like the mean and variance can be expressed as functions of the proportions p_J:

$$\mu = \sum_J p_J Y_J$$

$$\sigma^2 = \sum_J p_J (Y_J - \mu)^2.$$

A simple random sample from the population can be correspondingly expressed through the sample counts n_J at Y_J (most of these will be zero). If the sample size n is small compared to the population size N, (the alternative case is considered later) so that sampling with replacement accurately approximates sampling without replacement, the multinomial probability of the sample counts n_J is

$$\Pr(n_1, \ldots, n_D) = m(n;\, p_1, \cdots, p_D) = \frac{n!}{\prod_{J=1}^{D} n_j!} \prod_{J=1}^{D} p_J^{n_J},$$

where the factorial term gives the number of distinguishable arrangements of the sample values. Given the sample counts, the likelihood is the *multinomial likelihood*

$$L(p_1, \ldots, p_D) = \prod_{J=1}^{D} p_J^{n_J},$$

where the combinatorial term $n! / \prod_{J=1}^{D} n_j!$ is a known constant and is omitted from the likelihood in all the applications in this chapter and others. The *sequence* in which the observations are drawn (among the different sequences contributing to the combinatorial term) does not play any role in inference about the model parameters, or in model comparisons, though it may in *model checking* (Chapter 7).

Formally, we need to know the smallest and largest values Y_1 and Y_D of Y, and the number of distinct values D in the population, to be able to compute this likelihood, but for any unobserved values of Y_J the corresponding n_J is zero, so the likelihood can be reexpressed in terms of the p_j for only the observed Y_J. Thus the p_J for the unobserved Y_J do not contribute to the likelihood, and so these Y_J do not need to be known unless the prior gives them nonzero weight.

Maximizing the multinomial likelihood over the p_J for a fixed μ gives the "empirical" profile likelihood in μ (Owen 1988), discussed extensively in Owen (2001).

4.2.2 Dirichlet prior

A fully Bayesian analysis follows from the specification of a prior for the p_J; a very convenient choice is the natural conjugate Dirichlet prior, used by Hartley and Rao (1968), Ericson (1969), and Rubin (1981), which has density

$$\pi(p_1, \ldots, p_D) = C(a_1, \ldots, a_D) \prod_{J=1}^{D} p_J^{a_J - 1}$$

over the D-dimensional simplex $p_J > 0$, $\sum_{J=1}^{D} p_J = 1$, where $C(a_1, \ldots, a_D)$ is the normalizing constant:

$$C(a_1, \ldots, a_D) = \Gamma\left(\sum_1^D a_J\right) / \prod_1^D \Gamma(a_J).$$

The posterior distribution is again Dirichlet:

$$\pi(p_1, \ldots, p_D | y) = C(n_1 + a_1, \ldots, n_D + a_D) \prod_1^D p_J^{n_J + a_J - 1}.$$

Priors and posteriors for functions of the p_J follow automatically from the Dirichlet prior for the p_J: no additional prior specifications are necessary (for example, for the mean μ).

The Dirichlet is a special case of the *Dirichlet process prior* (Ferguson 1973); this was used by Binder (1982) for the more general case of finite population values which are arbitrary real numbers. However, for most real-sampled populations with fixed measurement or recording precision, the simpler equally spaced grid of population values is sufficient, and we restrict consideration to this case. The multinomial/Dirichlet model and prior have been proposed recently as the fundamental nonparametric distribution and prior model, in the work of Gutiérrez-Pena and Walker (2005) and Walker and Gutiérrez-Pena (2007).

Since even in large samples many of the positive values of n_J will be 1 or a small integer, the choice of prior is more important than usual in parsimonious parametric models. The effective information provided by the prior is easily seen from the form of the posterior: the sample counts n_J are augmented by the prior "weights" a_J. The "total prior weight" $a = \sum_J a_J$ augments the total sample weight $n = \sum_J n_J$.

Ericson (1969) considered the proper prior with $a_J = \epsilon_J$ with $\epsilon = \sum_J \epsilon_J$ "small," of the order of 1. He showed that many standard survey sampling results followed as limiting cases as $\epsilon \to 0$, though he expressed reservations about the properties of such an unrealistically "rough" prior.

Rubin (1981) introduced the term *Bayesian bootstrap* for posterior inference with the improper Haldane prior with $a_J = 0 \forall J$. This prior is used by Gutiérrez-Pena and Walker (2005) and Walker and Gutiérrez-Pena (2007). It leaves the total sample weight unchanged, but has the curious property that for any values Y_J not observed in the sample, the posterior distribution for the corresponding p_J has a nonintegrable spike at zero. This is equivalent to assigning zero prior probability to these unobserved values. The computation of the posterior distribution can then be restricted to the d observed distinct sample values y_J rather than the D distinct population values, a great saving. This saving is shared with Owen's empirical likelihood, and with frequentist bootstrapping: These procedures also depend only on the observed sample values and their sample frequencies.

The term "Bayesian bootstrap" comes from the analogy with the frequentist bootstrap, which resamples from the observed sample. The Bayesian bootstrap also uses only the observed sample, but it resamples from the *posterior distribution* of the *probabilities* attached to each observed value, rather than from the values themselves.

Rubin (1981, pp. 133–4) highlighted difficulties he saw with the Haldane prior approach:

> ... First, is it reasonable to use a model specification that effectively assumes all possible distinct values of [Y] have been observed?

* * *

> ... Second, even assuming all distinct values of [Y] have been observed, is it reasonable to assume a priori independent parameters, constrained only to sum to 1, for these values? If two values of [Y] are 'close', isn't it often realistic to assume that the associated probabilities of their occurrence should be similar? Shouldn't the parameters be smoothed in some way?

Banks (1988) took up these criticisms by developing a smoothing of the Dirichlet posterior: given the Haldane prior, he proposed generating a random value of p_J for each observed Y_J, and then spreading it uniformly over this Y_J and all unobserved values to the left of this Y_J down to the next observed value. In this way the posterior mass was spread over the whole sample range from $y_{(1)}$ to $y_{(n)}$, though in an ad hoc way.

We examine the behavior of the Haldane prior: an apparently unreasonable prior specification would be expected to perform poorly. Aitkin (2008) demonstrated the contrary with a simulation study of several methods. The study evaluated the frequentist performance of the Bayesian bootstrap relative to other frequentist procedures, following the precept that to be useful, Bayes procedures need to be *well calibrated* in the frequentist sense (Rubin 1987, p. 62). An important property of credible intervals provides this calibration in some models.

4.2.3 Confidence coverage of credible intervals

In large samples from parsimonious regular parametric models with flat priors and normal likelihoods, $100(1 - \alpha)\%$ central credible intervals or regions are identical to $100(1 - \alpha)\%$ likelihood-based central confidence intervals or regions. We illustrate with the one-dimensional case, though the results apply generally.

For a sample of size n from a model which gives a *normal likelihood* in the parameter θ:

$$L(\theta) = \exp\left\{-\frac{1}{2\sigma^2}(\theta - \hat{\theta})^2\right\},$$

with an improper flat prior $\pi(\theta) = c$, the posterior distribution of θ is $N(\hat{\theta}, \sigma^2)$, and so the 95% central credible interval for θ is $\hat{\theta} \pm 1.96\sigma$. If the likelihood in θ is normal for *all* data sets of size n in repeated sampling from the population, then $\hat{\theta}$ is normally distributed $N(\theta, \sigma^2)$, and the 95% credible interval is also a 95% confidence interval for θ.

The credible interval has the given credibility coefficient *without any assumption* about other unobserved data sets from the population, but the *calibration* of its confidence coverage depends on the assumption of a normal likelihood *in all other possible data sets of the same size* from the population. This is in general unverifiable without extensive simulation studies from *known* populations. Several such studies (for example, the Agresti and Min (2005) study of differences in binomial proportions) have concluded that the Jeffreys prior provides the best approximation of credible interval coverage to confidence coverage. Since the Jeffreys prior gives an improper posterior for samples with observed proportions of zero or one, it cannot be generally adopted for binomial data; we do not pursue this issue further as we concentrate on noninformative priors. A separate issue is a major difficulty with such comparisons: the confidence interval coverage itself varies in general with the true value of the parameter.

In any case, the multinomial model is nonparsimonious, and the sample size at each sample support point may be very small, so the large-sample result may not apply to credible intervals from the Bayesian bootstrap for derived parameters.

We first give a brief example of the income data given above.

4.2.4   Example – income population

For the simple random sample of 40 shown earlier, the sample mean is $\bar{y} = 67.1$ and the (unbiased) variance is $s^2 = 500.87$. Figure 4.1 shows the full income population as a maximum resolution histogram.

The survey-sampling large-sample 95% confidence interval for the mean is $\bar{y} \pm 1.96s/\sqrt{n}$, which is [60.1, 74.0]; this is nearly identical to the t-interval [59.9, 74.3], assuming a normal distribution for income. The design-based interval using the finite population correction of $(1 - 40/648) = 0.938$ gives the slightly shorter interval [60.6, 73.6]. If income could be assumed to be normally distributed, the equal-tailed 95% confidence interval based on the χ^2_{39} distribution for the income variance would be [336.1, 825.9].

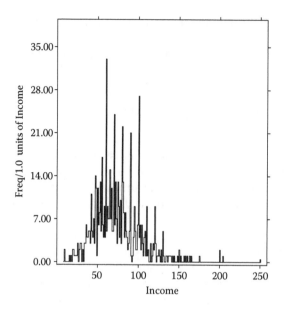

FIGURE 4.1
Income histogram.

For the Bayesian bootstrap analysis we tabulate the sample by the distinct values of Y in Table 4.2.

We first make an analogous notational change: since we use only the d ordered distinct sample values, we will denote them by y_j with sample frequencies n_j.

Simulation of the p_j, and therefore of any *marginal* function of the p_j, from the Dirichlet posterior with the Haldane prior is particularly simple: for a single simulation we generate $d = 30$ independent $U(0, 1)$ values U_j, transform them to d independent gamma variables G_j with parameters 1 and n_j, and then define $p_j = G_j / \sum_j G_j$. Repeating the simulations M times gives

TABLE 4.2

Income Data Tabulation

j	1	2	3	4	5	6	7	8	9	10
y_j	26	35	38	39	42	46	47	52	53	55
n_j	1	1	1	1	1	1	3	1	1	2
j	11	12	13	14	15	16	17	18	19	20
y_j	56	58	60	65	67	69	70	71	72	75
n_j	1	1	5	2	2	1	1	1	1	1
j	21	22	23	24	25	26	27	28	29	30
y_j	77	80	81	85	93	96	104	107	119	120
n_j	1	1	1	1	1	1	2	1	1	1

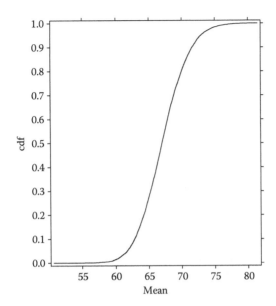

FIGURE 4.2

Posterior cdf: income mean.

M simulated values $p_j^{[m]}$ of the p_j, and hence M simulated values

$$\mu^{[m]} = \sum_j p_j^{[m]} y_j$$

from the marginal posterior distribution of μ.

We show in Figure 4.2, the posterior cdf of the mean μ from a simulation of size $M = 10,000$, and in Figure 4.3 a Gaussian kernel density estimate for μ using a bandwidth of 1.0, together with the simulated values.

(Approximate) percentiles of the posterior distribution can be read directly from Figure 4.2 (or from the list of ordered values). The posterior density in Figure 4.3 has only very mild skew. The simulation sample median is 67.0, and the simulation sample mean is 67.1, identical to the observed sample mean (Ericson 1969). The 95% equal-tailed credible interval for the mean is [60.6, 74.2]; it is slightly shorter than the t-interval and slightly asymmetric.

Simulation is not restricted to the mean μ – we can simulate *any* parametric function of the p_j – the variance or standard deviation and higher moments are just as simple.

Figure 4.4 shows the joint posterior scatter of the $M = 10,000$ values of μ and σ for the income sample, and Figure 4.5 shows the joint posterior scatter of the standardized third and fourth cumulants of the income distribution. The point (0,0) is in the extreme edge of the point scatter in Figure 4.5: there is no question that the population is both skewed and heavy-tailed.

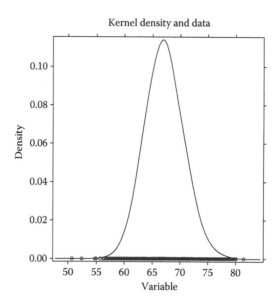

FIGURE 4.3
Posterior density: income mean.

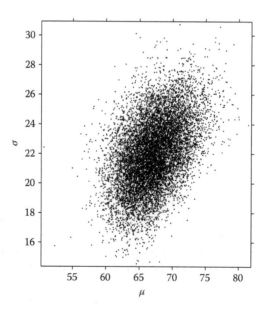

FIGURE 4.4
Joint posterior: income μ and σ.

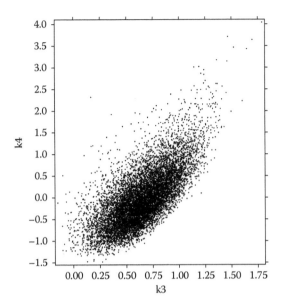

FIGURE 4.5
Joint posterior: income k3 and k4.

The 95% equal-tailed credible interval for the variance is [308.3, 708.0], substantially different from the normal-based interval of [336.1, 825.9], and the corresponding interval for the standard deviation is [17.6, 26.6]. The posterior distribution of *percentiles* can be simulated in the same way. We illustrate with the median and the 75th percentile.

The population median Y_{med} is defined as the largest value of Y such that $Pr[Y \leq Y_{med}] \leq 0.5$. Roughly speaking, the median is the value above which the cumulative distribution function of Y changes from less than (or equal to) 0.5 to greater than 0.5. This is easily simulated: from the mth draw of the d values $p_j^{[m]}$ we form the cumulative probabilities $c_j^{[m]} = \sum_{k=1}^{j} p_k^{[m]}$, and find the median value $Y_{med}^{[m]}$ for this draw; this is a draw from the posterior distribution of Y_{med}. Since the sample values of Y are discrete, the posterior distribution of the median is also discrete, on the same sample support. So the cdf will be constant between jumps at the observed support points, and the percentiles of the posterior distribution will be a discrete set.

Figure 4.6 shows the posterior distribution of the median from $M = 10,000$ draws, and Figure 4.7 shows the posterior distribution of the 75th percentile from the same set of draws.

The posterior median of the draws for the population median is 60, and for the 75th percentile is 75. The discrete posterior distributions of these percentiles are shown in Table 4.3.

We cannot set arbitrarily the credibility coefficients for credible intervals for percentiles because of the discreteness of the posterior distributions.

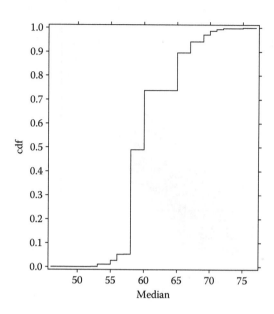

FIGURE 4.6
Posterior cdf: median.

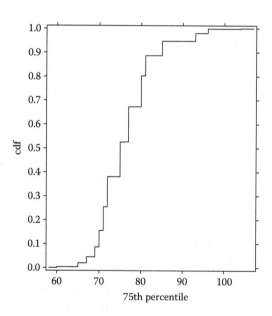

FIGURE 4.7
Posterior cdf: 75th percentile.

TABLE 4.3

Posterior cdfs of Median and 75th Percentile

Y	46	47	52	53	55	56	58	60
Median	0.0001	0.0005	0.0012	0.0112	0.0273	0.0529	0.4896	0.7380
75th							0.0003	0.0034

Y	65	67	69	70	71	72	75	77
Median	0.8956	0.9421	0.9708	0.9869	0.9935	0.9976	0.9995	1.000
75th	0.0194	0.0447	0.0865	0.1551	0.2544	0.3802	0.5258	0.6729

Y	80	81	85	93	96	104	107
75th	0.8025	0.8871	0.9472	0.9800	0.9986	0.9998	1.000

Approximate 96% credible intervals for the median and 75th percentile are [56, 70] and [67, 93], respectively. The true values are 70 and 90.

Since the multinomial model is nonparsimonious, the coverage properties of credible intervals in repeated sampling may not correspond to the credibility coefficient. Aitkin (2008) assessed this for credible intervals for the mean in a simulation study.

4.2.5 Simulation study

The study compared coverage of credible intervals for the StatLab income mean, based on the Bayesian bootstrap using the Haldane prior, with confidence intervals based on the gamma and normal distributions (the latter closely equivalent to the survey sampling interval). The full population cdf gave quite close agreement with a gamma distribution. As noted in Sections 4.2.2 and 4.2.3, the frequentist performance of the Bayesian bootstrap was evaluated relative to other frequentist procedures.

Aitkin (2008) drew 1000 random samples of size 40 from the StatLab boy population of $N = 648$, and constructed 80%, 90%, and 95% confidence intervals for the mean μ based on a normal income distribution (using t percentage points), and corresponding intervals based on a gamma income distribution with scale parameter r, from the sample mean and its estimated standard error $\bar{y}/\sqrt{n\hat{r}}$ using the MLE $\hat{r}$, and using t rather than normal percentage points. (This choice provides an approximate "small-sample" adjustment to the asymptotic normal percentage points which also allows a direct comparison with the normal income distribution intervals.) The StatLab income population was chosen for the simulation as it is a real one, and also has features which may be expected in other real populations, like irregularity, rounding, and preference for income values which are multiples of 5 and 10.

For each data sample, 10,000 frequentist bootstrap samples were drawn (by sampling with replacement) and the empirical 80%, 90%, and 95% bootstrap percentile intervals (equal-tailed) were constructed for the mean, and 10,000 Bayesian bootstrap samples were drawn from the posterior distribution of the mean based on the Haldane prior. Aitkin also extended the prior support to the full range of the observed sample, and used an Ericson-type Dirichlet

TABLE 4.4

Average Interval Length and Coverage, $n = 40$

Coefficient	Normal	Gamma	Boot	Haldane	Ericson
80%	11.90	11.68	11.54	11.26	11.32
90%	15.38	15.09	14.81	14.57	14.63
95%	18.46	18.12	17.64	17.49	17.55
c	0.799	0.789	0.777	0.770	0.767
80% lnc	0.079	0.085	0.092	0.097	0.112
rnc	0.122	0.126	0.131	0.133	0.121
c	0.890	0.892	0.879	0.876	0.872
90% lnc	0.038	0.039	0.046	0.047	0.056
rnc	0.072	0.069	0.075	0.077	0.072
c	0.945	0.941	0.933	0.927	0.925
95% lnc	0.017	0.020	0.023	0.030	0.034
rnc	0.038	0.039	0.044	0.043	0.041

prior with parameters $1/\ell$, where ℓ is the number of support points in the grid from $y_{(1)}$ to $y_{(n)}$. The last prior gives an equivalent prior weight of 1 compared to the sample weight of 40. From these samples the equal-tailed 80%, 90%, and 95% credible intervals for the mean were constructed.

In Table 4.4, we give in the first panel the average length of the intervals across the 1000 samples, and in the second panel the actual coverage (c) of the intervals, as well as the proportion of left (lnc) and right (rnc) noncoverage.

Apart from the Ericson prior, the table shows a consistent pattern: the interval lengths decrease slightly across the columns, and the coverages decrease slightly. The intervals based on the Ericson prior on the extended observed support behave qualitatively differently: compared with the Haldane prior, interval lengths increase but coverages *decrease*. Greater prior weight on the unobserved values accentuates this effect (results not shown): it further decreases the coverage and increases the length of the credible intervals.

Two criticisms can be made of the Ericson prior:

- It biases the posterior mean toward the sample median, which is inappropriate since the sample income distribution is clearly skewed.
- It cannot be assigned until the data are observed, so it is a postdata prior.

These criticisms illustrate the difficulty of the Dirichlet approach if it requires prior assignment to unobserved values – how is this to be done? However, this comparison shows that even a prior weight of 1, over a conservative range, compared to the sample weight of 40 results in poorer coverage and longer intervals than the Haldane prior. We now consider the comparison of the Haldane prior intervals with those from the other methods.

The pattern of shorter interval lengths with reduced coverage makes it difficult to compare the methods directly – would a method with lower coverage but shorter intervals than another method have the same, better, or worse coverage if the interval length were increased to match that of the other method?

We address this question by (frequentist) modeling of the coverage probabilities in a probit analysis with interval length as an explanatory variable; the interval methods are an explanatory factor which is tested for significance in the analysis. We regress the probits of the coverage probabilities against the interval lengths for the sample size of 1000, with method (5 levels) and nominal coverage (3 levels) as explanatory factors.

This analysis shows that coverage probability is very strongly determined by interval length; method and nominal coverage show no significant variation once interval length is included; the simple interval length model has a goodness-of-fit χ^2 value of 2.24 with 13 degrees of freedom.

Thus the methods are equivalent in coverage after adjustment for interval length: the t, model-based gamma, frequentist bootstrap, and Haldane prior *intervals of the same length perform equally well in coverage*, though the t and model-based gamma methods have the closest to nominal coverage.

The apparently "unreasonable" Haldane prior provides the best set of credible intervals: apparently more reasonable priors which do not exclude unobserved values perform less well than the Haldane prior.

4.2.6 Extensions of the Bayesian bootstrap

The above discussion of the Bayesian bootstrap analysis is limited to simple random sampling and moment and percentile parameters of the multinomial distribution of Y. In the remainder of this chapter we extend the Bayesian bootstrap approach in several directions. Section 4.3 extends the one-sample approach to sampling without replacement, using the approach of Hoadley (1969). Section 4.4 discusses regression models and gives an example supported by a simulation study. Section 4.5 extends the multinomial model to multiple subpopulations. This provides the analysis for stratified sampling in Section 4.6 and for cluster sampling in Section 4.7. Section 4.8 discusses a complex example of regression in a stratified and clustered sample. Section 4.9 has discussion and conclusions; it will be clear from the extensions that the Bayesian bootstrap approach can handle survey designs of considerable complexity.

4.3 Sampling without replacement

The multinomial likelihood construction in Section 4.1 is based on the assumption that the sample size n is small compared with the population size N. When this is not so, we need an appropriate construction of the likelihood. The probability that a sample of size n contains n_J of the N_J values of Y_J in the population is now the hypergeometric probability

$$\Pr(n_1, \ldots, n_D) = \left[\prod_{J=1}^{D} \binom{N_J}{n_J} \right] \bigg/ \binom{N}{n}.$$

Given the sample values $n_1, \ldots, n_D$, the likelihood in the parameters $N_1, \ldots, N_D$ is the *hypergeometric likelihood*

$$L(N_1, \ldots, N_D) = \prod_{J=1}^{D} \binom{N_J}{n_J},$$

where the known constant denominator is omitted, and the N_J must be greater than or equal to n_J. Since $\binom{N}{0} = 1$, the zero counts can again be omitted from the likelihood, which can be expressed in terms of only the observed sample counts n_j:

$$L(N_1, \ldots, N_d) = \prod_{j=1}^{d} \binom{N_j}{n_j}.$$

Thus the sample is again uninformative about the population counts at unobserved values of Y. (Note that if all $n_j = 1$, $L(N_1, \ldots, N_d) = N_1 \cdot N_2 \cdot \ldots \cdot N_d$, for the observed values $y_1, \ldots, y_d$. This *is* informative about these values of N_j.)

The population counts are not free parameters: they must satisfy $N_j \geq n_j$. Write $N_j^* = N_J - n_J$; we take the $N_j^* \geq 0$ to be the unknown parameters, with $\sum_{J=1}^{D} N_J^* = N - n = N^*$. The likelihood in the N_j^* is

$$L(N_1^*, \ldots, N_d^*) = \prod_{j=1}^{d} \binom{N_j^* + n_j}{n_j}.$$

For this form of likelihood there is no simple conjugate prior distribution for the N_j^*. Following Hoadley (1969), we embed the model in *two* levels of prior distribution.

Conditional on category proportions p_j, we treat the d unobserved population counts N_j^* as drawn from a multinomial distribution

$$m(N^*; p_1, \cdots, p_d) = \frac{N^*!}{\prod_{j=1}^{d} N_j^*!} \prod_{j=1}^{d} p_j^{N_j^*},$$

in which the probabilities p_j, conditional on the observed sample sizes n_j, have the Dirichlet distribution of Section 1:

$$\pi(p_1, \ldots, p_d \mid n_1, \ldots, n_d) = \frac{\Gamma(n)}{\prod_{j=1}^{d} \Gamma(n_j)} \prod_{j=1}^{d} p_j^{n_j - 1}.$$

Integrating out the p_j gives a *compound multinomial distribution* as the posterior distribution of the N_j^* given the n_j:

$$\Pr[N_1^*, \ldots, N_d^* \mid n_1, \ldots, n_d] = \int \cdots \int \Pr[N_1^*, \ldots, N_d^* \mid p_1, \ldots, p_d]$$

$$\cdot \Pr[p_1, \ldots, p_d \mid n_1, \ldots, n_d] \mathrm{d}p_1 \ldots \mathrm{d}p_{d-1}$$

$$= \int \cdots \int \frac{N^*!}{\prod_{j=1}^d N_j^*!} \prod_{j=1}^d p_j^{N_j^*}$$

$$\cdot \frac{\Gamma(n)}{\prod_{j=1}^d \Gamma(n_j)} \prod_{j=1}^d p_j^{n_j - 1} \mathrm{d}p_1 \ldots \mathrm{d}p_{d-1}$$

$$= c \cdot \frac{N^*!}{\prod_{j=1}^d N_j^*!} \prod_{j=1}^d \frac{\Gamma(N_j^* + n_j)}{\Gamma(N^* + n)}.$$

This distribution does not lend itself to direct simulation, but its integral derivation provides a very simple indirect sampling formulation for simulation of the mean μ (or other parameters):

- Generate M values $p_j^{[m]}$ of the p_j as in Section 4.2.4.
- From these, generate M values $N_j^{*[m]}$ from the multinomial distributions $m(N^*; p_1^{[m]}, \cdots, p_d^{[m]})$.
- Calculate the M values

$$\mu^{[m]} = \sum_{j=1}^d (N_j^{*[m]} + n_j) Y_j / \sum_{j=1}^d (N_j^{*[m]} + n_j).$$

This approach avoids completely the awkward form of the posterior in the N_j^*, and requires only the additional multinomial simulation step. An alternative approach to the unobserved N_j, not used in this book, is the Polya urn model (Ghosh and Meeden 1997, p. 42).

4.3.1 Simulation study

Aitkin (2008) replicated part of the simulation study in Section 4.2.5, with the same sample size from the StatLab population, to compare the posterior distributions of the mean for sampling with and without replacement. The study was restricted to just the two posteriors based on the same Haldane prior-based posterior for the p_j. Results are given in Table 4.5.

The intervals based on the hypergeometric likelihood are shorter, but have lower coverage, than those based on the multinomial likelihood. Adjusting again for interval length, the coverages are equivalent – the frequentist

TABLE 4.5

Average Credible Interval Length and
Coverage, $n = 40$

Coefficient	with Rep	without Rep
80%	11.34	10.99
90%	14.67	14.21
95%	17.61	17.06
c	0.756	0.741
80% lnc	0.126	0.134
rnc	0.118	0.125
c	0.858	0.846
90% lnc	0.069	0.077
rnc	0.073	0.077
c	0.923	0.913
95% lnc	0.036	0.041
rnc	0.041	0.046

deviance for the single-interval length model is 0.046 on 4 df. Recognizing the finiteness of the population does not bring increased precision in inference about its mean in this example; the sample fraction of $40/648 = 0.062$ is not sufficiently large to give the theoretical improvement.

4.4   Regression models

We want to relate an outcome variable Y to an explanatory variable X through a regression. We use an example from Royall and Cumberland (1981) for illustration: they discussed a finite population of 393 short-stay hospitals for which data were available on the number of patients Y discharged in one year and the number of hospital beds X in that year. The data came from the NCHS Hospital Discharge Survey, a national sample of short-stay hospitals with fewer than 1000 beds (Herson 1976). We treat this as the population for this study. A simple random sample of size $n = 32$ is shown in Figure 4.8 and the values of Y and X are given in Table 4.6.

We are interested in the total number of short-stay patients across the population, and we assume a simple proportionality of the number of such patients in each hospital to the number of beds in that hospital. We know from administrative records the number of beds X in each hospital and hence the total number of beds T_X in all the hospitals ($T_X = 107,956$), and draw an SRS of hospitals of size n, recording in each hospital the number of short-stay patients and the number of beds. From the sample data we want to estimate the total number T_Y of short-stay patients in all the hospitals.

Figure 4.8 shows that the variance of Y is clearly increasing with X, so the *ratio estimator* may be an appropriate choice in the survey sampling approach.

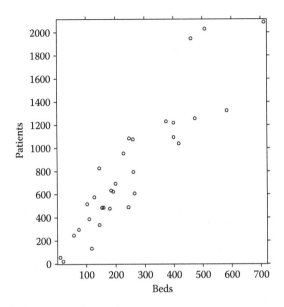

FIGURE 4.8
Patients and beds for hospital sample.

The ratio estimator is

$$\hat{T}_Y = \frac{\sum_{i=1}^n y_i}{\sum_{i=1}^n x_i} \cdot T_X$$

$$= \frac{\bar{y}}{\bar{x}} T_X$$

$$= \hat{B} T_X$$

TABLE 4.6

Number of Patients Y and Hospital Beds X

Y	1076	577	1258	134	795	1219	486	1095
X	260	128	474	118	261	400	154	400
Y	1040	297	22	625	955	1948	1084	57
X	418	74	20	192	228	461	247	10
Y	828	487	795	1326	2031	2089	518	695
X	145	159	261	584	509	712	103	200
Y	247	635	1231	609	337	490	389	479
X	57	185	374	265	145	244	110	180

where

$$\hat{B} = \sum_{i=1}^{n} y_i / \sum_{i=1}^{n} x_i.$$

From a model-based viewpoint, this estimator would be optimal (in the weighted least squares sense) under a model in which Y has mean BX and variance $\sigma^2 X$.

4.4.1 Design-based approach

Using the notation in Section 2.3, we index the population values by $I = 1, \ldots, N$, and denote the corresponding Y and X values by Y_I^* and X_I^*. We define the *population ratio regression coefficient* by

$$B = \frac{\sum_{I=1}^{N} Y_I^*}{\sum_{I=1}^{N} X_I^*} = \frac{T_Y}{T_X} = \frac{\mu_Y}{\mu_X},$$

and the population total T_Y estimate by $\hat{T}_Y = \hat{B} T_X$. Let Z_I be the sample selection indicators, with $Z_I = 1$ if population member I is included in the sample, and $Z_I = 0$ otherwise. We have

$$\hat{B} = \frac{\sum_{I=1}^{N} Y_I^* Z_I}{\sum_{I=1}^{N} X_I^* Z_I}.$$

The sampling distribution of this ratio is complicated by the appearance of the Z_I in both numerator and denominator. Exact results are not available, but an approximate variance for $\hat{B}$ is $(1 - n/N)s_e^2/(n\mu_X^2)$ (Lohr 1999, p. 68), where

$$s_e^2 = \sum_{i=1}^{n} (y_i - \hat{B} x_i)^2 / (n-1),$$

and this can be used to construct approximate confidence intervals for B, and hence for T_Y. An alternative robust *sandwich* variance estimate is obtained by replacing the normal model variance estimate $\sigma^2 / \sum_{i=1}^{n} x_i$ by

$$\text{Var}[\hat{B}] = \sum_{i=1}^{n} \text{Var}[Y_i] / \left(\sum_{i=1}^{n} x_i\right)^2 \doteq \sum_{i=1}^{n} (y_i - \hat{B} x_i)^2 / \left(\sum_{i=1}^{n} x_i\right)^2 \doteq s_e^2/(n\bar{x}^2),$$

where the variance model is not assumed to be correct.

4.4.2 Bayesian bootstrap approach

We follow the same approach as in the simple mean model. The population consists of N pairs Y_I^*, X_I^*. We tabulate them conceptually into the D *distinct* pairs (Y_J, X_J) with frequency N_J. The probability that a randomly drawn sample value gives the pair (Y_J, X_J) is $p_J = N_J/N$. Our interest is not in the p_J but in the ratio regression coefficient

$$B = \sum_{J=1}^{D} p_J Y_J / \sum_{J=1}^{D} p_J X_J.$$

We draw a random sample of size n (we assume for the moment with replacement) and obtain counts n_J for the distinct values. The likelihood of the sample is as before (omitting the known constant)

$$L(\mathbf{p}) = \prod_{J=1}^{D} p_J^{n_J}.$$

We use the Haldane Dirichlet D(**0**) prior with $a_J = 0$ for all J, giving the Dirichlet posterior D(**n**), now defined on the d distinct values in the observed support:

$$\pi(p_1, \ldots, p_d \mid \mathbf{y}) = \frac{\Gamma(n)}{\prod_{j=1}^{d} \Gamma(n_j)} \prod_{j=1}^{d} p_j^{n_j - 1}.$$

We draw $M = 10,000$ random values $p_j^{(m)}$ of the p_j on the observed support, and compute the 10,000 values of $B^{(m)}$. The (approximate) 95% central credible interval for B is computed from the 250th and 9750th ordered values of $B^{(m)}$.

For the hospital example, the 10,000 values of B generated from the posterior distribution of the p_j give a median of 3.200 and a central 95% credible interval of [2.917, 4.515] (the population value is 2.966). The corresponding credible interval for T_Y is [314,908, 379,465]. The true value is 320,159. The posterior cdf and density estimate for B are shown in Figures 4.9 and 4.10; those for T_Y are just rescaled.

Note that any other function of the p_J could be simulated in the same way; for example, if it were clear that the variance of Y was proportional to X^2 rather than X, while the mean was linear in X, the estimate of B^* would be

$$\hat{B}^* = \sum_i (y_i / x_i),$$

implying a population definition of

$$B^* = \sum_J p_J Y_J / X_J,$$

whose posterior distribution could be simulated at the same time as that of B.

4.4.3 Simulation study

Aitkin (2008) used the hospital population in a simulation study of the performance of the survey sampling estimate and approximate confidence interval,

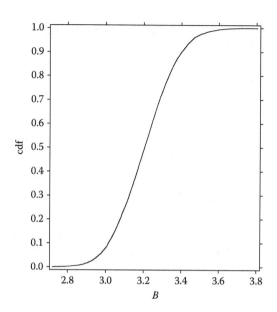

FIGURE 4.9
Posterior cdf of ratio regression coefficent B.

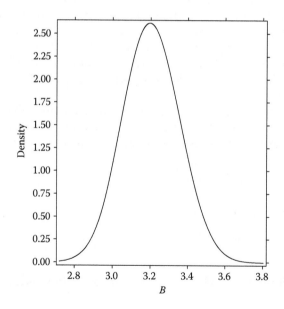

FIGURE 4.10
Posterior density of ratio regression coefficient B.

the confidence interval from the normal model using the information-based standard error, the bootstrap confidence interval, and the credible interval using the Haldane prior.

He generated 1000 random samples of size $n = 50$ from the hospital population, and for each sample constructed the 80%, 90%, and 95% confidence intervals for the population ratio regression coefficient by five methods:

- The MLE assuming a normal model with variance proportional to X (the MLE is identical to the ratio estimate) and with standard error from the normal model information matrix
- The MLE assuming a normal model with variance proportional to X but with *robust* "sandwich" standard error from the normal model information matrix and the squared residuals
- The sample survey ratio estimate, with approximate standard error from the sampling distribution of the Z_I
- The bootstrap central percentile interval from 10,000 bootstrap samples
- The central credible interval from 10,000 samples from the posterior distribution of B

These are given in Table 4.7.

Apart from the normal intervals with the information-based standard error, all the methods performed very similarly, with slight undercoverage at the higher confidence levels. Relative coverage is not related to interval length in this example, and the only significant effect in the probit analysis of coverage proportions is the lower coverage of the first normal method. The information-based standard error appears to be too small, probably a consequence of the variance of Y not being proportional to X.

TABLE 4.7

Average Interval Length and Coverage, $n = 50$

Coefficient	Normal	Sandwich	Survey	Boot	Haldane
80%	0.2814	0.3337	0.3362	0.3300	0.3187
90%	0.3632	0.4307	0.4339	0.4232	0.4109
95%	0.4353	0.5162	0.5200	0.5036	0.4919
c	0.724	0.794	0.802	0.796	0.784
80% nlc	0.149	0.113	0.117	0.118	0.122
nrc	0.127	0.093	0.081	0.086	0.094
c	0.835	0.886	0.880	0.882	0.879
90% nlc	0.092	0.064	0.075	0.070	0.076
nrc	0.073	0.050	0.045	0.048	0.055
c	0.904	0.943	0.937	0.937	0.931
95% nlc	0.050	0.033	0.044	0.040	0.042
nrc	0.046	0.024	0.019	0.023	0.027

4.4.4   Ancillary information

It might appear that the Bayesian analysis could be strengthened by using the additional information in the hospital bed population. We know the *marginal* proportions of hospitals with exactly X beds. If the marginal *sample* proportions departed from these, then it would appear that we should adjust the posterior distribution of T_Y (which was just that for B scaled by T_X) by scaling-up the predictions for each X_J by the actual population proportions at these X_J. However, the scaling by T_X already achieves this (since T_X incorporates these population proportions), so the marginal proportions of hospitals at each bed number cannot provide more information. This result follows also from incomplete data theory (Little and Rubin 1987): if the unobserved Y are "missing at random" (that is, selection into the sample is not based on Y) then the full information about B is contained in the observed sample pairs (y_j, x_j), and the additional observed X_J provide no further information about B, and hence about T_Y.

4.4.5   Sampling without replacement

The analysis and simulations above assume that sampling is with replacement. For sampling without replacement, we may simply adapt the hypergeometric analysis in Section 4.3 to the regression model. As before the population consists of N pairs Y_I^*, X_I^*. We tabulate them conceptually into the D *distinct* pairs (Y_J, X_J) with frequency N_J. The probability that a randomly drawn sample value gives the pair (Y_J, X_J) is $p_J = N_J/N$. The ratio regression coefficient is now expressed as

$$B = \sum_{J=1}^{D} N_J Y_J \Big/ \sum_{J=1}^{D} N_J X_J.$$

We draw a random sample of size n, now without replacement, and obtain counts n_J for the distinct values. Given the sample values $n_1, \ldots, n_D$, the likelihood in the parameters $N_1, \ldots, N_D$ is the hypergeometric likelihood

$$L(N_1, \ldots, N_D) = \prod_{J=1}^{D} \binom{N_J}{n_J},$$

which is expressed in terms of only the observed sample counts n_j:

$$L(N_1, \ldots, N_d) = \prod_{j=1}^{d} \binom{N_j}{n_j}.$$

As before, write $N_J^* = N_J - n_J$, with $\sum_{J=1}^{D} N_J^* = N - n = N^*$. The likelihood in the N_j^* is

$$L(N_1^*, \ldots, N_d^*) = \prod_{j=1}^{d} \binom{N_j^* + n_j}{n_j}.$$

Conditional on category proportions p_j, we treat the d unobserved population counts N_j^* as drawn from a multinomial distribution

$$m(N^*; p_1, \cdots, p_d) = \frac{N^*!}{\prod_{j=1}^{d} N_j^*!} \prod_{j=1}^{d} p_j^{N_j^*},$$

in which the probabilities p_j, conditional on the observed sample sizes n_j, have the Dirichlet distribution of Section 2.3:

$$\pi(p_1, \ldots, p_d \mid n_1, \ldots, n_d) = \frac{\Gamma(n)}{\prod_{j=1}^{d} \Gamma(n_j)} \prod_{j=1}^{d} p_j^{n_j-1}.$$

Integrating out the p_j gives a *compound multinomial distribution* as the posterior distribution of the N_j^* given the n_j:

$$\Pr[N_1^*, \ldots, N_d^* \mid n_1, \ldots, n_d] = \int \cdots \int \Pr[N_1^*, \ldots, N_d^* \mid p_1, \ldots, p_d]$$

$$\cdot \Pr[p_1, \ldots, p_d \mid n_1, \ldots, n_d] \, dp_1 \ldots dp_{d-1}$$

$$= \int \cdots \int \frac{N^*!}{\prod_{j=1}^{d} N_j^*!} \prod_{j=1}^{d} p_j^{N_j^*}$$

$$\cdot \frac{\Gamma(n)}{\prod_{j=1}^{d} \Gamma(n_j)} \prod_{j=1}^{d} p_j^{n_j-1} \, dp_1 \ldots dp_d$$

$$= c \cdot \frac{N^*!}{\prod_{j=1}^{d} N_j^*!} \prod_{j=1}^{d} \frac{\Gamma(N_j^* + n_j)}{\Gamma(N^* + n)}.$$

We use the simple indirect sampling formulation for simulation of the ratio regression coefficient:

- Generate M values $p_j^{[m]}$ of the p_j as in Section 2.3.
- From these, generate M values $N_j^{*[m]}$ from the multinomial distributions $m(N^*; p_1^{[m]}, \cdots, p_d^{[m]})$.
- Calculate the M values

$$B^{[m]} = \sum_{j=1}^{d} \left(N_j^{*[m]} + n_j\right) Y_j \Big/ \sum_{j=1}^{d} \left(N_j^{*[m]} + n_j\right) X_j.$$

TABLE 4.8

Average Credible Interval Length and Coverage, $n = 40$

Coefficient		with Rep	without Rep
80%		0.3551	0.3362
90%		0.4578	0.4337
95%		0.5481	0.5192
	c	0.769	0.743
80%	lnc	0.125	0.136
	rnc	0.106	0.121
	c	0.867	0.850
90%	lnc	0.071	0.081
	rnc	0.062	0.069
	c	0.929	0.911
95%	lnc	0.036	0.046
	rnc	0.035	0.043

4.4.6   Simulation study

Aitkin (2008) replicated part of the simulation study in Section 4.3.3, but with sample size 40 from the hospital's population, to compare the posterior distributions of the ratio regression coefficient for sampling with and without replacement. The study was restricted to just the two posteriors based on the same Haldane prior-based posterior for the p_j. Results are given in Table 4.8.

The hypergeometric intervals are shorter, but have lower coverage, as in the single-sample case. The probit analysis of actual coverage with length, method, and nominal coverage as explanatory variables shows that the simple interval length model is sufficient to describe the results: the deviance of this model is 0.104 on 4 df, so there is no improvement in coverage from the hypergeometric likelihood. The theoretical gain in precision is again not visible with a sampling fraction of 0.10.

4.5   More general regression models

The approach in this section can be readily extended to general regression models. For complex models we adopt the "working model" language of Valliant et al. (2000, p. 50), in which the "working" probability model leads to an optimal estimator under the model, which is then used without the working model being assumed to hold.

Suppose that a working model has $E[Y|\mathbf{x}] = \mathbf{B}'\mathbf{x}$, $\text{Var}[Y] = \sigma^2$, leading to the usual least squares estimate $(X'X)^{-1}X'\mathbf{y}$ of $\mathbf{B}$. This can be immediately treated in the same Bayesian way, expressed by definition as $\mathbf{B} = (X'PX)^{-1}X'P\mathbf{y}$, where P is a diagonal weight matrix of the population proportions at each support point in the $(Y, \mathbf{x})$ space. We simulate M values $p_j^{[m]}$ from the posterior Dirichlet distribution of the p_j, giving the M values

$\mathbf{B}^{[m]} = (X'P^{[m]}X)^{-1}X'P^{[m]}\mathbf{y}$. This will in general require M matrix inversions of the weighted SSP matrix. Nonconstant variance models can be easily incorporated.

The ability to use standard software for the Dirichlet analysis with an additional weight vector greatly extends the generality of the Bayesian bootstrap analysis. We illustrate this with the analysis of a complex example in Section 4.8.

4.6 The multinomial model for multiple populations

Consider a population of size N which is made up of S subpopulations indexed by $s = 1, \ldots, S$, with N_s members and proportion $p_s = N_s/N$ in subpopulation s. A response variable of interest Y takes values in the full population. As for the case of a single population, we conceptually tabulate the full population by the *distinct* values $Y_1 < \ldots < Y_J < \ldots < Y_D$. In subpopulation s the proportions of the subpopulation at the values Y_J are denoted by $p_{sJ} = N_{sJ}/N_s$, where N_{sJ} is the number of members at Y_J in subpopulation s. We do not assume that the proportions p_{sJ} are related across subpopulations: the set of multinomial distributions is completely general. However, we use the observed support across all subpopulations to define a *minimally informative* Dirichlet prior, since in the individual subpopulations not all the values of the observed support appear. We use the minimally informative prior with prior weights $1/d$ on the d distinct pooled sample values $y_1 < \ldots < y_d$ of y; this gives an effective prior sample weight of 1. The subpopulation means and variances of Y are

$$\mu_s = \sum_{J=1}^{D} p_{sJ} Y_J, \quad \sigma_s^2 = \sum_{J=1}^{D} p_{sJ}(Y_J - \mu_s)^2.$$

(In Aitkin 2008 a different Haldane prior was used for each sample, defined on the subsamples from each population, but the support information from the pooled samples is relevant for posterior inference about all the sets of subpopulation proportions, so the minimally informative prior links the sample information in a natural way. The results for the example below are essentially identical to those in Aitkin 2008, as the prior sample size is 1 compared to the pooled sample size of 30.)

We draw a random sample of size n_s from the sth subpopulation, with total sample size $n = \sum_{s=1}^{S} n_s$, and obtain n_{sj} sample values at y_j in the sth subpopulation; some of these will be zero. The *sample fraction* π_s is the proportion n_s/N_s drawn from the sth subpopulation.

The subpopulation sample means and variances are

$$\bar{y}_s = \sum_{j=1}^{d} n_{sj} y_{sj}/n_s, \quad s_s^2 = \sum_{j=1}^{d} n_{sj}(y_{sj} - \bar{y}_s)^2/(n_s - 1).$$

TABLE 4.9

Subpopulation Structure

	Subpopulation	Proportion	Sample Fraction	Mean	Variance
Population	s	$p_s = N_s/N$		μ_s	σ_s^2
sample	s	n_s/n	$\pi_s = n_s/N_s$	$\bar{y}_s$	s_s^2

We consider only the case when the sample fractions are small, so that sampling with and without replacement are equivalent. Large sample fractions require the hypergeometric likelihood rather than the multinomial; the approach of Section 4.3 can be adapted to the more general case here.

The overall population mean is

$$\mu = \sum_{s=1}^{S} N_s \mu_s / N = \sum_{s=1}^{S} p_s \mu_s,$$

and the overall sample mean is

$$\bar{y} = \sum_{s=1}^{S} n_s \bar{y}_s / n.$$

These are conveniently summarized in Table 4.9.

We make use of this structure for two different types of sampling.

4.7   Complex sample designs

4.7.1   Stratified sampling

Stratified sampling is designed to reduce variability in estimation due to known population heterogeneity – the population is made up of homogenous subpopulations with substantial differences in mean and/or variance among them. If some of the subpopulations are small, a simple random sample may miss them completely, or give only small subsamples from them. Stratified sampling sometimes *oversamples* small strata to give comparable sample sizes from all strata – the assessment of strata differences is most precise, for a fixed total sample size, when the strata sample sizes are proportional to the strata variances (and so are equal if the strata variances are equal).

We now change notation, using the usual subscript h to denote *stratum*. For a single response variable Y, we wish to estimate the stratum means μ_h and the overall population mean, allowing for the different sample fractions $\pi_h = n_h/N_h$ in the different strata. For Bayesian inference about the individual μ_h, we proceed as for the single population mean in Section 4.1, but with the minimally informative Dirichlet prior. We draw M values $p_{hj}^{[m]}$ in stratum h from the posterior Dirichlet distribution of the p_{hj} on the observed support

y_j in that stratum, using the minimally informative prior (the same for each stratum), and map these into M values

$$\mu_h^{[m]} = \sum_{j=1}^{d} p_{hj}^{[m]} y_j$$

of the posterior distribution of μ_h. Then for posterior inference about $\mu = \sum_{h=1}^{H} p_h \mu_h$, we simply combine the M simulated values of μ_h:

$$\mu^{[m]} = \sum_{h=1}^{H} p_h \mu_h^{[m]}$$

to give M values from the posterior distribution of μ. We postpone an example to Section 4.8.

Note that the sample fractions π_h do not play any role in this analysis: they affect only the precision of the posteriors for μ_h through the actual sample sizes in each stratum. No *weighting* by the inverse of the sample fractions is needed, only weighting by the actual subpopulation proportions.

It is possible, though unlikely, for a small stratum to be missed completely even in stratified sampling, if it is decided that this stratum is of no interest. Then no information can be gained about this stratum mean, so formally no information can be gained about the overall population mean either. In practice this amounts to redefining the mean of the *population of interest*, to be the weighted mean of the means of the strata of interest.

4.7.2   Cluster sampling

Cluster sampling has a similar formal structure to stratification, but the population parameters of interest are different. Cluster sampling is a form of two-stage sampling, in which the population is divided into clusters which are defined by geographic contiguity or other similarities, which make units sampled within the same cluster more homogeneous than those sampled from different clusters. Clustering frequently reduces sampling costs compared with simple random sampling.

The two-stage design selects clusters at random according to a sample design, and samples units within clusters according to a second sample design (sometimes a full sample of all units in the clusters).

The analysis in cluster sampling allows for the greater homogeneity *within* clusters than that *among* clusters, and this is naturally represented through *variances*.

We now change notation, using the subscript c to represent cluster identification; the design has C clusters. We adapt Table 4.9 to represent clustering in Table 4.10:

TABLE 4.10

Clustering in Subpopulation Structure

	Cluster	Proportion	Sample Fraction	Mean	Variance
Population	c	$p_c = N_c/N$		μ_c	σ_c^2
cluster	c	n_c/n	$\pi_c = n_c/N_c$	$\bar{y}_c$	s_c^2

The overall population mean is $\mu = \sum_{c=1}^{C} p_c \mu_c$, and the overall population variance is

$$\sigma^2 = \sum_{c=1}^{C} \sum_{J=1}^{D} N_{cJ}(Y_J - \mu)^2/N$$

$$= \sum_{c=1}^{C} \sum_{J=1}^{D} p_{cJ} N_c (Y_J - \mu_c + \mu_c - \mu)^2/N$$

$$= \sum_{c=1}^{C} N_c \left[\sigma_c^2 + \sum_{J=1}^{D} p_{cJ}(\mu_c - \mu)^2 \right]/N$$

$$= \sum_{c=1}^{C} p_c \left[\sigma_c^2 + (\mu_c - \mu)^2 \right]$$

$$= \sigma_W^2 + \sigma_A^2$$

where N_{cJ} is the number of cluster c values of Y equal to Y_J, $p_{cJ} = N_{cJ}/N_c$, $\sigma_W^2 = \sum_c p_c \sigma_c^2$ is the (average) *pooled within-cluster variance* and $\sigma_A^2 = \sum_c p_c(\mu_c - \mu)^2$ is the *among-cluster variance*.

A common form of cluster sampling uses *probability proportional to size* (PPS) sampling – we sample clusters at the same sampling rate, so the larger clusters have the larger sample sizes. This means that small clusters may not be sampled at all, leading to the same situation described for omitted strata – no information can be obtained about the omitted cluster means and variances. This leads to effectively the same definition of the population of interest, as that which has been sampled. An alternative is to assume a formal parametric model, which disguises this situation, though it clearly cannot provide any additional information about the cluster means and variances for the unsampled clusters.

An extreme case, pointed out by Jon Rao (personal communication), is where the sampled clusters are *fully enumerated*, leading to complete information about the proportions p_{cJ} in the sampled clusters, but no information in the unsampled clusters. This would lead to a single point posterior for the overall population mean and variance based on only the sampled clusters. This case does not present a practical difficulty, as PPS sampling would lead to the opposite data pattern: the larger clusters would not be fully enumerated, and the omitted clusters would be small. We assume, in the examples discussed in the following paragraphs, that either all clusters are sampled, or that the population definition refers to the clusters which are sampled.

The posterior distributions of both the *variance components* σ_W^2 and σ_A^2 can be simulated directly from their definitions in terms of the cluster means, variances, and proportions. As for the case of stratified sampling, denote the sample data count at the distinct pooled sample values y_j from cluster c by n_{cj}. The cluster population proportion at y_{cj} is p_{cj}, and given the sample data, we simulate M values $p_{cj}^{[m]}$ from the posterior Dirichlet distribution of the p_{cj} with the minimally informative prior defined over the pooled support y_j as in Section 4.6.1. From these we compute the M values $\mu_c^{[m]}$, $\sigma_c^{[m]2}$, $\mu_c^{[m]}$, $\sigma_A^{[m]2}$, and $\sigma_W^{[m]2}$ from their definitions.

An important point here is that the cluster sizes N_c in the population do not need to be known for this analysis, nor the total population size N: only the proportions p_c are used, and these are based on the sample proportions at each observed value.

We illustrate with a small example from Box and Tiao (1973, p. 246), a designed experiment in which five samples were randomly chosen from six batches of raw material, and a single laboratory determination made (of the yield of dyestuff in grams of standard color) on each sample. This example is artificial for population survey sampling, but our aim is to show how variance components are estimated.

Box and Tiao (1973) gave the details of the Bayesian treatment of the normal variance component model, and the joint posterior distribution of the "among-batch" and "within-batch" variance components. In Box and Tiao's Bayesian analysis the within-batch variances are assumed to be the same across batches; we relax this assumption.

The data are given in Table 4.11, with the batch mean and (unbiased) variance.

For each batch c, we generate $M = 10,000$ random values $p_{jc}^{[m]}$ of the p_{jc} for the observed values y_{jc} in that batch, and substitute them in the various means and variances for each c, and the variance components. The posterior distributions for the batch means and variances are shown in Figures 4.11 and 4.12.

They are very diffuse, a consequence of the small sample size in each batch. (The batch can be identified in the figures by matching the sample mean to the

TABLE 4.11

Dyestuff Data

Batch	1	2	3	4	5	6
	1545	1540	1595	1445	1595	1520
	1440	1555	1550	1440	1630	1455
	1440	1490	1605	1595	1515	1450
	1520	1560	1510	1465	1635	1480
	1580	1495	1560	1545	1625	1445
Mean	1505.0	1528.0	1564.0	1498.0	1600.0	1470.0
variance	3975.0	1107.5	1442.5	4720.0	2500.0	962.5

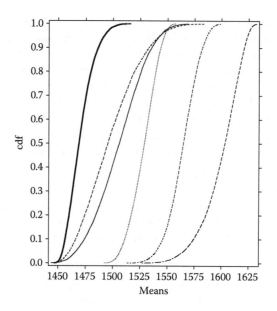

FIGURE 4.11
Posteriors: batch means.

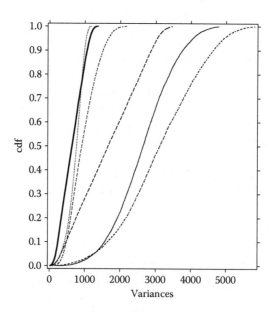

FIGURE 4.12
Posteriors: batch variances.

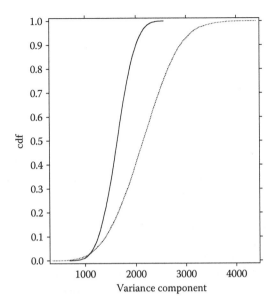

FIGURE 4.13
Posteriors: batch variance components.

posterior median.) The sample means differ considerably, and there is some sample variance heterogeneity – the largest variance ratio between batches is 4.90. The posterior distributions of the among-batch and pooled within-batch variance components are shown in Figure 4.13.

4.7.3 Shrinkage estimation

The variance components are widely used in *small-area estimation*, in which the estimation of an area mean is improved by "borrowing strength" from the other area means through their variation as measured by the among-area variance component.

In fully (normal) model-based inference a *shrinkage estimator* of an area mean may be superior to the simple area sample mean, if the area sample size is small. In the normal two-level model

$$y_{jc} \mid \mu_c \sim N(\mu_c, \sigma_c^2),$$

$$\mu_c \sim N(\mu, \sigma_A^2)$$

it follows immediately that

$$\bar{y}_c \mid \mu_c \sim N(\mu_c, \sigma_c^2/n_c),$$

$$\mu_c \mid \bar{y}_c \sim N(\mu + w_c(\bar{y}_c - \mu), \sigma_A^2(1 - w_c))$$

where

$$w_c = \frac{n_c \theta_c}{1 + n_c \theta_c}, \quad \theta_c = \frac{\sigma_A^2}{\sigma_c^2}.$$

We write the posterior expectation of μ_c as

$$\mu_{cPE} = w_c \bar{y}_c + (1 - w_c)\mu.$$

This is widely used as a *shrinkage estimator* of the area mean. The difficulty with this estimator in frequentist theory (apart from the assumption of normality) is how to specify correctly its variability; in the fully normal model-based analysis *empirical Bayes* estimators are widely used, with MLEs replacing the unknown variance component parameters and overall mean, but the variability of the resulting shrinkage estimator is very difficult to establish; further, the posterior *variance* of the μ_c is widely ignored.

The same Dirichlet posterior analysis provides the inference about the μ_c, correctly adjusted for parameter uncertainty. We first substitute the simulated values above into the variance component ratio and the means, using $\mu^{[m]} = \sum p_c \mu_c^{[m]}$, giving

$$\theta_c^{[m]} = \frac{\sigma_A^{[m]2}}{\sigma_c^{[m]2}}$$

$$w_c^{[m]} = \frac{n_c \theta^{[m]}}{1 + n_c \theta^{[m]}}$$

$$\mu_{cPE}^{[m]} = w_c^{[m]} \mu_c^{[m]} + \left(1 - w_c^{[m]}\right)\mu^{[m]}$$

$$\mu_{cPV}^{[m]} = \sigma_A^{[m]2}\left(1 - w_c^{[m]}\right)$$

where μ_{cPV} is the posterior variance of μ_c. If we can assume that

$$\mu_c \mid \bar{y}_c \sim N\big(\mu + w_c(\bar{y}_c - \mu), \sigma_A^2(1 - w_c)\big)$$

then for each m we draw one random value $\mu_c^{[m]}$ from $N(\mu_{cPE}^{[m]}, \mu_{cPV}^{[m]})$. These values allow for *all* the uncertainty in the parameters, *and* for the variance of the posterior distribution of μ_c.

This is one of the great strengths of the Bayesian analysis: the simulation variability in the population proportions p_j is *propagated* throughout the subsequent functions of these parameters. We illustrate this with the Box and Tiao example in Section 4.7.2.

Figure 4.14 shows the posterior distributions (as cdfs), for each batch, of the "fixed effect" mean μ_c of the batch (*without* using the batch random effect distribution – solid curves) and of the batch random effect (dashed curves), derived as described above from $M = 10,000$ samples.

Surprisingly, the random effect posterior distributions are more diffuse than those for the "fixed effects" – it appears that incorporating the additional information has *decreased*, rather than *increased*, the precision of inference!

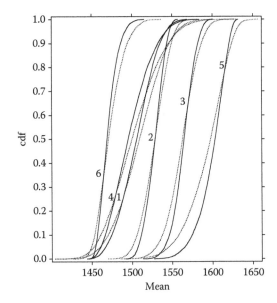

FIGURE 4.14
Posterior and fixed effect means: all batches.

There are two reasons for this result. First, we have not assumed that the batch variances are homogeneous. In a frequentist analysis, this assumption is made routinely; without it, the batch variance for each batch has four degrees of freedom instead of the 24 of the pooled within-batch variance. As a result, *all* of the batch means, variances, and variance component ratio are based on very small samples and are very imprecise. The additional information provided by the among-batch variance component (which is itself based on only five degrees of freedom) does not overcome this imprecision.

Second, we are not assuming a normal (or any other) parametric distribution for the dyestuff yield, and so the multinomial distributional model in each batch is based on only five observed values.

The homogeneity of variance assumption is important for inference about the batch means: If we use the *average batch variance* instead of the individual batch variances, the random effect batch posteriors (not shown) are *more precise* than the fixed effect posteriors, and also show shrinkage toward the overall population mean. This is the usual conclusion from empirical Bayes analyses, but its validity depends strongly here on the homogeneity of variance assumption.

4.8   A complex example

We conclude this chapter with an analysis of the Labor Force survey data from Valliant et al. (2000, App. B.5) given by Aitkin (2008). The sample of 478 individuals is stratified in three strata, and clustered in 115 clusters within strata, with an average of four individuals per cluster. We illustrate the general approach with a main-effect regression of weekly wage on sex, age, and hours worked, allowing for the stratification and clustering.

Since the (stratum, cluster) cells hold only four cases each on average, we assume a constant variance of wage across these cells. We comment on this assumption below.

We index the data by (i, c, h) for person i in cluster c and stratum h, and write y_{ich} for the weekly wage of person i in cluster c and stratum h, a_c and b_h for the random cluster and fixed stratum effects, and $\mathbf{x}_i$ for the explanatory variables on person i. We adopt the working model

$$E[y_{ich} \mid a_c] = \beta' \mathbf{x}_i + a_c + b_h$$

$$\mathrm{Var}[y_{ich} \mid a_c] = \sigma^2$$

$$E[a_c] = 0$$

$$\mathrm{Var}[a_c] = \sigma_A^2$$

$$\mathrm{Cov}[y_{ich}, y_{i'c'h'}] = \delta(c, c') \sigma_{A'}^2$$

where $\delta(c, c') = 1$ if $c = c'$ and zero otherwise. These are the usual assumptions of the two-level cluster random effect model with fixed stratum effects. If in addition the cluster and wage variables were assumed to be normally distributed, the optimal estimates of β and the stratum effects would be the ML estimates from the two-level normal variance component model. These can be expressed as generalized least squares estimates: writing $\mathbf{b}$ for the vector of stratum effects and $Z = [X, B]$ for the design matrix of explanatory variables and stratum effects, the MLEs are the solutions of

$$Z' V Z[\beta, \mathbf{b}] = Z' V \mathbf{y}$$

where V is the block-diagonal covariance matrix of the observations. This solution can be obtained from any standard two-level maximum likelihood model program. We adopt the ML estimators of fixed effects and variance components as defining the population parameters for the Bayesian bootstrap analysis, but without the assumption of normality.

We express the posterior distributions of the population parameters – regression coefficients, stratum effects, and variance components – in terms of the posterior Dirichlet distributions of the probabilities at each sample point in each cluster within stratum, based on the number (2–5) of persons in each cluster. The full simulation procedure is surprisingly simple:

- From the observations within each cluster, construct the Dirichlet posterior, with the Haldane prior, of the probabilities p_{jc} on the observed support y_{jc} within that cluster.

- Draw M values $p_{jc}^{[m]}$ from the posteriors of the p_{jc}.

- Using the $p_{jc}^{[m]}$ as explicit *weights* for each observation y_{jc} in cluster c, carry out M weighted ML fits of the y_{cj} to the explanatory variables, to obtain parameter estimates $\beta^{[m]}$, $\mathbf{b}^{[m]}$, $\sigma^{[m]2}$ and $\sigma_A^{[m]2}$.

These M values provide the required posterior distributions of the parameters.

For the conventional two-level normal model analysis we assume the wage variance is constant. For the main effect model of age, sex, hours worked, and stratum, the MLEs and standard errors (omitting those for strata) are given below.

Maximum Likelihood Estimates and Standard Errors, Wage Example

	Age	Sex	Hours	Sigma	Sigma_A
MLE	1.988	-107.8	7.061	136.7	76.76
SE	0.125	3.46	0.154		

The Bayesian bootstrap analysis with $M = 10,000$ gives posterior distributions (not shown) for all the parameters which are indistinguishable from normal, apart from slightly heavier tails for the among-cluster variance component. The posterior means and standard deviations for the parameters are shown below.

Posterior Means and Standard Deviations, Wage Example

	Age	Sex	Hours	Sigma	Sigma_A
MEAN	1.792	-115.2	7.514	137.2	77.01
SD	0.412	10.6	0.497	4.03	8.28

The variance component estimates are very close; the other parameter estimates are less close but similar, as might be expected from a weighted analysis, but their precisions are very different – the posterior SDs are 3 to 4 times as large as the SEs. The reason for this is very clear – a graph of cluster sample variance against mean shows that the wage variance increases with mean, so the constant variance model gives a compromise variance which misrepresents the nature of the variability. The Bayes analysis gives a variance estimate which is almost the same, but the effect of the weighting is similar to that of "sandwiching" the variance estimates in a frequentist analysis which allows for variance heterogeneity: the model uncertainty, in *both* distributional *and* variance terms, is allowed for by the Dirichlet distributions in each cluster.

We emphasize that *any* package which can both simulate gamma random variables and fit two-level models could be used for this analysis; we do not give specific package code.

4.9   Discussion

The Bayesian bootstrap approach to finite population analysis is quite general. It accepts the survey sampling axiom of not assuming the correctness of a full parametric model for the population form of the response variable, but it nevertheless provides full information about the defined population parameters through the multinomial likelihood and the noninformative Haldane Dirichlet prior. The prior restricts analysis to the observed support, in the same way as maximum empirical likelihood and the frequentist bootstrap. This approach, like parametric model-based approaches, does not use the sampling distributions of the sample-selection indicators, but unlike parametric model approaches, does not require the examination or validation of the parametric model by residual examination.

It may seem surprising that the Bayesian bootstrap analysis is equivalent to a series of M weighted maximum likelihood analyses with randomly varying weights; the variation among the M resulting parameter realizations provides the full posterior distribution of these parameters, in a (model) distribution-free way, since it depends only on the primitive multinomial model, which is not a model assumption which can be contradicted by the data.

The ability to use standard software packages (provided these allow for weights) is a particularly useful feature of the Bayesian bootstrap analysis. Running 10,000 regression analyses may appear computationally intensive, especially for generalized linear mixed models (GLMMs), but since the weights are varying only randomly, the MLEs can be used as starting values for each analysis, and so convergence of each of the M model-fitting steps can be much faster than for the ML analysis itself.

This analysis has some similarities to the empirical likelihood analysis of Owen (2001), but the latter depends on asymptotic frequentist likelihood theory for the calibration of empirical likelihood confidence intervals and regions. As noted in the examples in this chapter, the credible intervals tend to undercover in small samples (as do likelihood-based confidence intervals in small samples from some parametric models), but this calibration is based only on the equivalence in larger samples of credible and likelihood-based confidence intervals; the credible intervals always have their usual Bayesian interpretation.

The disadvantages of the Bayesian bootstrap approach are shared with survey sampling analysis, empirical likelihood, and bootstrapping: without an explicit response probability model there is no optimal choice for population or regression parameters. The decision to use (say) the ratio estimator is not based on data properties, except insofar as the form of the variance function can be assessed from data plotting or residual examination. Different choices of the power of X parameter in the variance function lead to different estimators, but there is no obvious way of choosing which is more appropriate, since the multinomial likelihood is a function only of the population proportions at each support point, and not explicitly of the variance parameter.

Different variance parameters provide different population regression coefficient definitions, for all of which the multinomial likelihood and Dirichlet posterior provide credible intervals for the regression coefficients, but we are unable to compare these coefficients (and the implicit variance functions which justify them) through the model likelihoods, since they depend on the same unconstrained multinomial parameters.

A more refined examination of the variance form (for example, the choice of the most appropriate value of the power parameter of X) requires an explicit parametric model for the response, or else a constrained multinomial likelihood, to provide different likelihoods for different models which can then be compared. A similar problem occurs with the variance homogeneity assumption: this is a *model constraint* on the multinomial probabilities in each batch which is difficult to implement in a Bayesian framework. (An approach to this problem has been proposed by Carota 2008).

Thus the Bayesian bootstrap approach is not completely general, but within its limitations optimal Bayes procedures are available and fruitful, and readily computable with general software.

A major question is whether, and when, this analysis is necessary because the usual parametric assumptions are not met. This question is addressed by diagnostic model comparisons, to which we return in Chapter 7.

5

Regression and Analysis of Variance

We begin with the normal regression model.

5.1 Multiple regression

We extend the development in Aitkin et al. (2005). Consider the normal regression model with n observations on a response Y and a vector $\mathbf{x}$ of p explanatory variables and 1, with the model

$$Y \mid \mathbf{x} \sim N(\mu, \sigma^2), \ \mu = \beta'\mathbf{x}.$$

Our aim is to assess the important variables through a series of model comparisons expressed in terms of partitions $\beta' = (\beta_1', \beta_2')$ and hypotheses $H_1 : \beta_2 = 0$ in the reduced model. This is a traditional approach in frequentist analysis and in many Bayesian treatments, though as noted earlier it is largely dismissed by Gelman et al. (2004), because of their dissatisfaction with point null hypotheses.

The likelihood for the full model is

$$L(\beta, \sigma) = \frac{1}{(\sqrt{2\pi})^n \sigma^n} \exp\left\{-\frac{1}{2\sigma^2}\left[RSS + (\beta - \widehat{\beta})' X' X (\beta - \widehat{\beta})\right]\right\}.$$

where RSS is the residual sum of squares from the full model. In our analysis, we take flat priors on β and $\log \sigma$ to give the usual joint posterior distribution, with

$$\beta \mid \mathbf{y}, \sigma \sim N(\widehat{\beta}, \sigma^2 (X'X)^{-1}), \ RSS/\sigma^2 \mid \mathbf{y} \sim \chi^2_{n-p-1}.$$

However, this is not the prior used in Bayes analyses which rely on Bayes factors for model comparison, because it is improper. Users of Bayes factors have therefore to specify a proper informative prior, and (at least) examine the effect of this specification on the model comparisons.

A common "minimally informative" prior is *Zellner's g-prior* (Zellner, 1986):

$$\beta \mid \sigma^2 \sim N(\beta_p, g\sigma^2(X'X)^{-1}),$$

where β_p and g are prior hyper parameters to be specified. Since at least some of the components of β are expected to be zero, it is common for β_p to be taken as zero; with a large g the prior precision will be small compared to the likelihood.

The effect of this prior is to *shrink the posterior distribution of β uniformly* (in the parameters) *toward zero*:

$$\beta \mid \sigma, \mathbf{y} \sim N\left(\frac{g}{1+g}\widehat{\beta}, \ \frac{g}{1+g}\sigma^2(X'X)^{-1}\right).$$

The extent of the shrinkage depends on the value of g; as $g \to \infty$ the shrinkage goes to zero and the diffuse prior result is recovered. We do not use or discuss this prior further.

Consider the null hypothesis, $H_1 : \beta_2 = \mathbf{0}$, for some partition, against the full model alternative H_f that β_2 has no zero components – that the full model is needed. The likelihood ratio and deviance difference are

$$LR_{1f} = \frac{L(\beta_1, \mathbf{0}, \sigma)}{L(\beta, \sigma)}$$

$$D_{1f} = \frac{1}{\sigma^2}\left[RSS_1 - RSS_f + (\beta_1 - \widetilde{\beta}_1)'X_1'X_1(\beta_1 - \widetilde{\beta}_1) - (\beta - \widehat{\beta})'X'X(\beta - \widehat{\beta})\right]$$

where RSS_f is the residual sum of squares from the full model, RSS_1 is the residual sum of squares from the reduced model using only $\mathbf{x}_1$, X_1 is the partition of X corresponding to β_1, and $\widetilde{\beta}_1$ is the MLE of β_1 in the reduced model.

The posterior distribution of DD_{1f} or LR_{1f} is easily simulated, by making M random draws $\sigma^{[m]}$ from the marginal χ^2_{n-p-1} posterior distribution of RSS_f/σ^2, and for each m, one random draw $\beta^{[m]}$ from the conditional normal posterior $N(\widehat{\beta}, \sigma^{[m]2}[X'X]^{-1})$ given the $\sigma^{[m]}$. The residual sums of squares from the models are known, and the quadratic forms in β are evaluated from the given MLEs and X matrices.

We use a well-known data set – the gas consumption data of Henderson and Velleman (1981), which has observations on the fuel consumption, expressed in miles per (U.S.) gallon, of 32 cars with 10 design variables on the engine and transmission. Paralleling backward elimination methods in frequentist theory, we examine the strength of evidence for the various models in the backward elimination sequence. The approach does not depend on the choice of variables – any submodel can be compared with the full model in the same way.

We follow the backward elimination sequence of Aitkin et al. (AFHD, 2008, p. 151), using log(mpg) as the response variable and the explanatory variables, listed in order of backward elimination: c, drat, s, t, log(disp), cb, g, and log(hp). The corresponding t-statistics for these variables are -0.082, -0.320, -0.340, -0.461, -0.723, -1.070, 1.052, and -4.372. Elimination ceased with a final model using log(wt) and log(hp).

In the backward elimination sequence in AFHD the sums of squares of eliminated variables were pooled with the error sum of squares from the full model, so the degrees of freedom of the *t*-statistics changed at each step.

In the analysis below we maintain the posterior distribution of β and σ from the full model – this gives a more realistic picture of the information about many parameters from the small sample.

We show in Figures 5.1 through 5.8 the posterior distributions of D, based on 1000 simulations, for the successive omitted partitions corresponding to

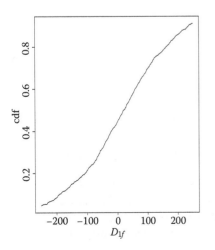

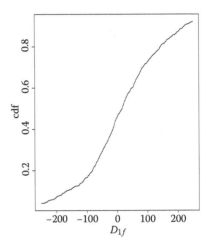

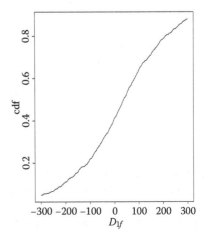

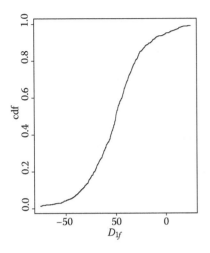

FIGURE 5.1
Step 1: c elimination.

FIGURE 5.2
Step 1: c elimination.

FIGURE 5.3
Step 3: s elimination.

FIGURE 5.4
Step 4: t elimination.

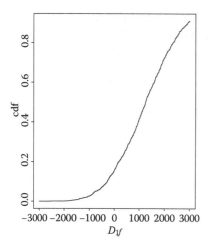

FIGURE 5.5
Step 5: log(disp) elimination.

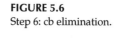

FIGURE 5.6
Step 6: cb elimination.

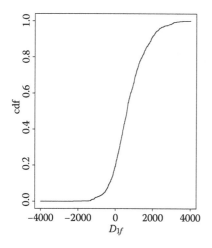

FIGURE 5.7
Step 7: g elimination.

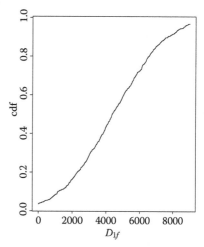

FIGURE 5.8
Step 8: log(hp) elimination.

the backward elimination t-statistics: {c}, {+ drat}, {+ s}, {+ t}, {+ log(disp)}, {+ cb}, {+ g}, {+ log(hp)}.

The distributions are all remarkably diffuse, with very large variances, reflecting the very small degrees of freedom of the residual sum of squares. The empirical tail probabilities that $D < 4.605$, that is, that $LR > 0.1$ (weak evidence against the null hypothesis of zero regression coefficients), are given in Table 5.1.

Despite this diffuseness, the message is very clear: the early distributions have large probabilities for $D < 4.605$, around 0.5 for the first four variables

TABLE 5.1

$Pr[D < 4.605]$ for Subset Elimination

Step	Variable Omitted	$Pr[D < 4.605]$
1	c	0.585
2	drat	0.473
3	s	0.458
4	t	0.420
5	log(disp)	0.159
6	cb	0.111
7	g	0.196
8	log(hp)	0.036

eliminated; this drops to around 0.16 at step 5 but increases again to around 0.2 in step 7. At step 8 the distribution changes drastically, with a tail probability below 4.605 of only 0.036. These results are completely consistent with the backward elimination t-statistics.

5.2 Nonnested models

This approach is easily extended to nonnested models. We illustrate with the well-known `trees` data set from the *Minitab Handbook* (Ryan et al., 1976). Aitkin et al. (2009) gave a discussion of the trees data in which different families of models were used to represent the regression of usable wood volume V on the height H and diameter D of the tree, for a sample of 31 black cherry trees. Models considered were

- 1: $V^{1/3} \sim N(\beta_0 + \beta_1 H + \beta_2 D, \sigma^2)$
- 2: $V \sim N((\beta_0 + \beta_1 H + \beta_2 D)^3, \sigma^2)$
- 3: $\log V \sim N(\beta_0 + \beta_1 \log H + \beta_2 \log D, \sigma^2)$
- 4: $V \sim N(\exp(\beta_0 + \beta_1 \log H + \beta_2 \log D), \sigma^2)$
- 5: $\log V \sim N(\beta_0 + 1 \times \log H + 2 \times \log D, \sigma^2)$

The first two models linearize the regression by a cube root transformation of V or a cube root link function, the third and fourth by a log transformation of V or a log link function (both with log-transformed explanatory variables). None of these models is nested in any other, and all have the same number of parameters. Model 5 is a special case of Model 3, a "solid body" model with $\beta_1 = 1$ and $\beta_2 = 2$, which includes the cone and the cylinder; the frequentist deviance increases by only 0.002 under this constraint.

In AFHD the models are compared through their frequentist deviances, but there is no distributional calibration of the differences:

- 1: 133.76
- 2: 143.21

- 3: 132.20
- 4: 142.44
- 5: 132.20

The asymptotic distributions for deviances may apply here: we check whether the likelihood distributions for the normal models 1 and 3 agree with their asymptotic forms. These follow easily from the normal regression model results in Section 5.1:

$$-2\log L(\beta, \sigma) = n\log 2\pi + n\log \sigma^2 + RSS/\sigma^2 + (\beta - \hat{\beta})'X'X(\beta - \hat{\beta})/\sigma^2$$
$$= -2\log L(\hat{\beta}, \hat{\sigma}) + W - n\log W - n(1 - \log n) + V,$$

where $W \sim \chi^2_{n-p}$ independently of $V \sim \chi^2_p$. Since the distributions for models 1 and 3 have the same $p = 3$ and $n = 31$, they have the same shape, but are anchored from the slightly different frequentist deviances. They are shown in Figures 5.9 and 5.10 (solid curves), with the asymptotic distributions given by $-2\log L(\hat{\beta}, \hat{\sigma}) + V$ (dotted curves).

The asymptotic curves are not very close to the exact normal model curves – the sample size is not large enough for the asymptotic distribution to hold for the three-parameter models. Figure 5.11 shows the normal deviance curves for Models 1 (solid), 3 (dotted), and 5 (dashed).

The differences between models 1 and 3 are quite small, and the plot of their deviance differences (Figure 5.12) shows that the empirical posterior

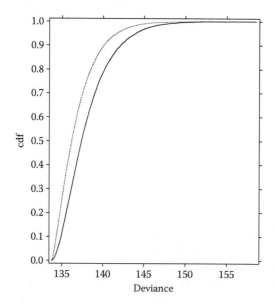

FIGURE 5.9
Normal and asymptotic deviances: Model 1.

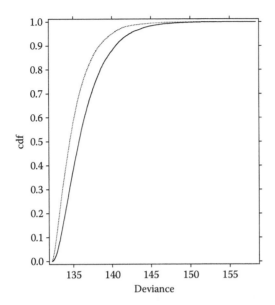

FIGURE 5.10
Normal and asymptotic deviances: Model 3.

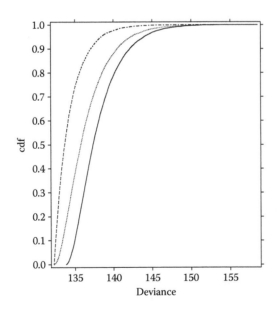

FIGURE 5.11
Models 1, 3, and 5.

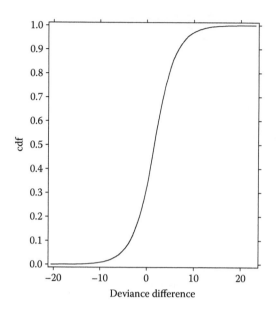

FIGURE 5.12
Model 1–3.

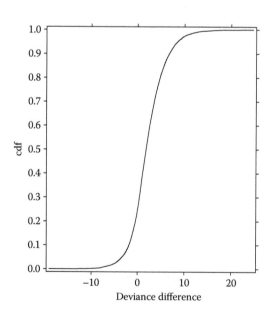

FIGURE 5.13
Model 3–5.

probability that the deviance difference is negative (in favor of Model 1) is 0.279. There is no clear preference between these models.

Comparing the constrained Model 5 with Model 3, the deviance plot in Figure 5.11 shows a much greater separation, because of the steeper slope of the χ_1^2 distribution. The normal deviance differences between Models 3 and 5 are shown in Figure 5.13.

The empirical probability of a negative difference (Model 5 is better) is 0.247, so (not surprisingly) the constrained model is preferred, but not very strongly.

Comparing the link transformed models with the transformed response models is less straightforward, because the link transformation models do not have normal likelihoods in the parameters. If the asymptotic deviance distribution results held for the link transformed models (which is less likely because of the nonlinear regressions in their parameters), it would follow immediately that the location shift in the frequentist deviances – of the order of 10 – would give almost no overlap in the shifted χ_3^2 distributions, and the link transformations would be ruled out as possible models.

6

Binomial and Multinomial Data

6.1 Single binomial samples

For the binomial case, we observe r successes in n trials with success probability θ. The likelihood is

$$L(\theta \mid r, n) = \binom{n}{r} \theta^r (1 - \theta)^{n-r}.$$

The conjugate beta(a, b) prior is

$$\pi(\theta) = \frac{1}{B(a, b)} \theta^{a-1} (1 - \theta)^{b-1};$$

this is proper for $a > 0$, $b > 0$. For this prior, the posterior is beta($r + a$, $n - r + b$); credible intervals for θ are available from the beta distribution percentiles. Common non- or weakly informative prior distributions for θ are versions of the conjugate prior:

- The uniform prior – beta(1,1): $\pi(\theta) = 1$
- The Jeffreys prior – beta($\frac{1}{2}$, $\frac{1}{2}$): $\pi(\theta) = \frac{1}{\pi} \theta^{-1/2} (1 - \theta)^{-1/2}$
- The Haldane prior – beta(0,0): $\pi(\theta) = c \cdot \theta^{-1} (1 - \theta)^{-1}$

The Haldane prior is uniform on the logit scale; this prior is *improper* and gives *improper posteriors* if $r = 0$ or n. The Haldane prior is therefore unsuitable when modeling rare-event data with small probabilities. The Jeffreys prior has infinite density at $\theta = 0$ and 1, so when $r = 0$ or n this prior *strongly reinforces* the mode in the likelihood at the boundary. It thus reinforces the data evidence for an extreme value of θ. This may not be desirable when modeling rare events. We use the uniform prior throughout this book for binomial data.

For a point null hypothesis $H_1 : \theta = \theta_1$ compared to a general alternative $H_2 : \theta \neq \theta_1$, the null model deviance is $D_1 = -2[r \log \theta_1 + (n - r) \log(1 - \theta_1)]$, and the deviance difference is

$$D_{12} = D_1 + 2[r \log \theta + (n - r) \log(1 - \theta)].$$

This is easily simulated: we make M draws $\theta^{[m]}$ from the beta posterior and substitute them in the deviance difference to give M draws from the deviance difference posterior:

$$D_{12}^{[m]} = D_1 + 2[r \log \theta^{[m]} + (n - r) \log(1 - \theta^{[m]})].$$

These draws can be mapped into draws from the likelihood ratio, and for specified model prior probabilities, into draws from the posterior distribution of the posterior probability of the null hypothesis. We give an example below.

6.1.1 Bayes factor

A direct evaluation of the Bayes factor is straightforward. For a general proper Beta prior with parameters a and b, the Bayes factor is the ratio:

$$
\begin{aligned}
BF_{12} &= \frac{L_1}{\binom{n}{r} \int \theta^r (1 - \theta)^{n-r} \pi(\theta) d\theta} \\
&= \frac{B(a, b)\theta_1^r (1 - \theta_1)^{n-r}}{\int \theta^{r+a-1}(1 - \theta)^{n-r+b-1} d\theta} \\
&= \frac{B(a, b)\theta_1^r (1 - \theta_1)^{n-r}}{B(r + a, n - r + b)}.
\end{aligned}
$$

A proper prior is, however, required for the existence of the Bayes factor, with $a > 0$, $b > 0$. The Haldane prior therefore cannot be used to compute a Bayes factor, though the posterior distribution of the likelihood ratio is well-defined for this prior so long as r, $n - r > 0$.

6.1.2 Example

We repeat in shortened form the Stone example from Section 2.8.3, as it highlights the gross difference which can occur between the Bayes factor and the posterior distribution of the likelihood with large samples and flat priors.

The example is of a physicist running a particle-counting experiment who wishes to identify the proportion θ of a certain type of particle. He has a well-defined scientific (null) hypothesis H_1 that $\theta = 0.2$, precisely. There is no specific alternative. He counts $n = 527, 135$ particles and finds $r = 106, 298$ of the specified type. What is the strength of the evidence against H_1?

The binomial likelihood function

$$L(\theta) = \binom{n}{r} \theta^r (1 - \theta)^{n-r} \doteq L(\hat{\theta}) \exp\left\{-\frac{(\theta - \hat{\theta})^2}{2SE(\hat{\theta})^2}\right\}$$

is maximized at $\theta = \hat{\theta} = 0.201652$ with standard error $SE(\hat{\theta}) = 0.0005526$. The standardized departure from the null hypothesis is

$$z_1 = (\theta_1 - \hat{\theta})/SE(\hat{\theta}) = 0.001652/0.0005526 = 2.9895,$$

with a two-sided p-value of 0.0028. The maximized likelihood ratio is $L(\theta_1)/L(\hat{\theta}) = 0.01146$.

The physicist uses the uniform prior $\pi(\theta) = 1$ on $0 < \theta < 1$ under the alternative hypothesis, and computes the Bayes factor

$$B_{12} = L(\theta_1)/\int_0^1 L(\theta)\pi(\theta)d\theta.$$

The denominator is

$$\bar{L} = \binom{n}{r}\int_0^1 \theta^r(1-\theta)^{n-r}d\theta$$
$$= \binom{n}{r}B(r+1, n-r+1)$$
$$= 1/(n+1),$$

which does not depend at all on the number of specified particles, only the total number of particles counted. The Bayes factor is then

$$BF_{12} = L(\theta_1)/\bar{L}$$
$$= 8.27$$

as derived in Section 2.8.3, giving a posterior probability of the null hypothesis, for equal prior probabilities, of 0.892.

Thus the p-value and Bayes factor are in clear conflict. However, the posterior distribution of θ is *not* in conflict with the p-value, since the posterior probability that $\theta > 0.2$ is $\Phi(2.990) = 0.9986 = 1 - p/2$. Any Bayesian using the uniform prior must have a very strong posterior belief that the true value of θ is larger than 0.2. Equivalently, the 99% HPD interval for θ is

$$\theta \in \hat{\theta} \pm 2.576SE(\hat{\theta}) = (0.20023, 0.20308)$$

which is identical to the 99% confidence interval, and excludes θ_1.

So the use of the Bayes factor leads to an evidential conclusion about the null hypothesis which is inconsistent with the posterior distribution using the same prior. However, the value of the Bayes factor is very sensitive to the length of the prior interval for θ. If instead of the full interval $(0,1)$ we take the flat prior interval of width $\pm 4SEs$ around the MLE (which covers the θ range over which the likelihood is appreciable), this has no effect on the posterior, but a *dramatic* effect on the Bayes factor, scaling it *down* by the ratio of the new interval length to the old length (1), that is, by 8×0.0005526, giving a new Bayes factor of 0.0364. Now the evidence is strongly *against* the null hypothesis, consistent with the credible interval conclusion.

The reader may argue that this approach violates temporal coherence – we must specify the prior *before* seeing the data, not after. (This violation of a Bayesian axiom is quite widespread, however, in "tuning the prior to the data" – choosing hyperparameters for priors in MCMC runs which place the parameters in the region of high likelihood, to avoid starting the chain in areas of flat likelihood which would slow or prevent convergence of the chain.)

The point is simply that, as is well known, the Bayes factor is very sensitive to the values of the hyperparameters defining the proper prior, in this case the length of the interval over which the flat prior is nonzero. With increasing sample size the likelihood becomes very sharply peaked, and so the integration of this sharply peaked function with respect to a uniform weight function over a fixed range produces a very small integrated likelihood. In this example the maximized likelihood is reduced to the integrated likelihood by the factor 721.94, giving a minute integrated likelihood against which the very small value of $L(\theta_1)$ looks large.

6.2   Single multinomial samples

The multinomial distribution and conjugate Dirichlet prior were discussed extensively in Chapter 4. Our interest here is not in "continuous" response values recorded on a grid, but essentially discrete categorical (or ordinal) values of a response variable. Given a random sample of size n with counts $n_1, \ldots, n_r$ in r response categories with corresponding category probabilities $p_1, \ldots, p_r$, the multinomial probability of this outcome is

$$m(\{n_i\} \mid n, \{p_i\}) = m(n_1, \ldots, n_r \mid n, p_1, \ldots, p_r) = \frac{n!}{\prod_{i=1}^{r} n_i!} \prod_{i=1}^{r} p_i^{n_i}.$$

The conjugate Dirichlet prior is $D(\{p_i\} \mid \{a_i\})$, with

$$\pi(p_1, \ldots, p_r \mid a_1, \ldots, a_r) = \frac{\Gamma(\sum_{i=1}^{r} a_i)}{\prod_{i=1}^{r} \Gamma(a_i)} \prod_{i=1}^{r} p_i^{a_i},$$

and the posterior is $D(\{p_i\} \mid \{n_i + a_i\})$. Posterior sampling proceeds similarly to the beta, but as the Dirichlet is multidimensional an indirect generation is used, as in Chapter 4: we make M random draws $g_i^{[m]}$ from the set of r independent gamma distributions with mean 1 and scale parameters $n_i + a_i$, sum across the r distributions and scale by the sum, to give

$$\{p_i^{[m]}\} = \left\{ g_i^{[m]} / \sum_{j=1}^{r} g_j^{[m]} \right\}.$$

We now consider two-way contingency tables, but first discuss a two-way table which reduces to a single-sample modeling problem.

6.3 Two-way tables for correlated proportions

The data in Table 6.1 come from Irony et al. (2000, their Example 2) and have been reanalyzed in Consonni and La Rocca (2008). They are artificial and are expressed in terms of a survey of 100 individuals expressing support (Yes/No) for the president, before and after a presidential address.

Denoting No by 1 and Yes by 2, we can write the observed table counts as $\mathbf{n} = (n_{11}, n_{12}, n_{21}, n_2)$, with the corresponding multinomial proportions $\pi = (\pi_{11}, \pi_{12}, \pi_{21}, \pi_2)$, with $\sum_j \sum_k \pi_{jk} = 1$.

The question of interest is whether there has been *change* in support between the surveys. If no *overall* change has occurred, then the switching proportions π_{12} and π_{21} must be equal, to a common π_S. We want to assess the evidence for the hypothesis of equality H_1 against the alternative hypothesis H_2 of a change.

6.3.1 Likelihood

The likelihood under the alternative hypothesis (omitting the canceling combinatorial constant) is

$$L_2(\pi_{11}, \pi_{12}, \pi_{21}, \pi_{22}) = \prod_j \prod_k \pi_{jk}^{n_{jk}},$$

and that under the null hypothesis is

$$L_1(\pi_{11}, \pi_S, \pi_{22}) = \pi_{11}^{n_{11}} \pi_S^{n_{12}+n_{21}} \pi_2^{n_{22}},$$

and the likelihood ratio is

$$LR_{12} = \pi_S^{n_{12}+n_{21}} / [\pi_{12}^{n_{12}} \pi_{21}^{n_{21}}].$$

The main-diagonal terms in the likelihood cancel, and so inference depends only on the off-diagonal counts, as in most frequentist (for example, the McNemar test) and Bayesian analyses.

We express the parametrization of the switching categories under the null hypothesis in terms of the alternative, by $\pi_S = (\pi_{12} + \pi_{21})/2$ (corresponding to the maximum likelihood estimate); in this parametrization the likelihood

TABLE 6.1

Support for the President

		After		
		No	Yes	Total
	No	20	21	41
Before	Yes	9	50	59
	Total	29	71	100

ratio is

$$LR_{12} = [(\pi_{12} + \pi_{21})/2]^{n_{12}+n_{21}}/[\pi_{12}^{n_{12}}\pi_{21}^{n_{21}}].$$

This is effectively a binomial likelihood ratio: write $\theta = \pi_{12}/(\pi_{12} + \pi_{21})$, $m = n_{12} + n_{21}$, and $r = n_{12}$, then

$$LR_{12} = \frac{(1/2)^m}{\theta^r(1-\theta)^{m-r}},$$

which is the likelihood ratio for comparing a symmetric binomial $\theta = 1/2$ with the general binomial $\theta \neq 1/2$, in m Bernoulli trials with r successes. The corresponding deviance difference is

$$D_{12} = 2[m\log 2 + r\log\theta + (m-r)\log(1-\theta)].$$

The posterior probability of the null hypothesis, given equal prior probabilities on the two hypotheses, is

$$\Pr[H_1 \mid, \mathbf{n}] = LR_{12}/[1 + LR_{12}].$$

6.3.2   Bayes factor

A direct evaluation of the Bayes factor is straightforward. As in Section 6.1, for the general beta prior with parameters a and b, the Bayes factor is

$$B_{12} = \frac{B(a,b)\theta_1^r(1-\theta_1)^{n-r}}{B(r+a, n-r+b)}.$$

Consonni and La Rocca (2008) gave a Bayes factor analysis which used an intrinsic prior depending on a prior sample size as a fraction q of the observed sample size.

6.3.3   Posterior likelihood ratio

The frequentist p-value for this example is 0.026 from the likelihood ratio test and 0.028 from the Pearson X^2 test, suggesting quite strong evidence against the null hypothesis. The posterior probability of the null hypothesis from the Bayes factor with the Jeffreys prior is 0.370, and for the uniform prior is 0.292. The posterior probability of the null hypothesis from the Consonni and La Rocca Bayes factor varied from 0.24 to 0.29 as q varied from 0 to 1. These suggest rather weak evidence against the null hypothesis.

   We apply the posterior likelihood ratio approach, with the three common conjugate priors: Haldane, Jeffreys, and uniform. For each prior we draw 10,000 values $\theta^{[m]}$ from the corresponding posterior beta distribution of θ, and substitute them in the likelihood ratio, deviance difference and posterior probability of H_1. Figure 6.1 shows the three posterior distributions for the deviance difference, and Figure 6.2 shows the posterior distributions for the

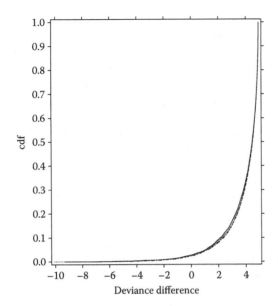

FIGURE 6.1
Deviance difference: Consonni and La Rocca example.

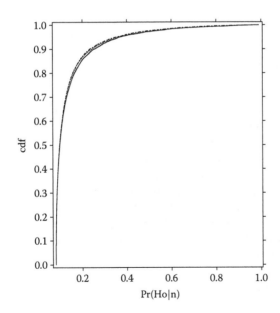

FIGURE 6.2
Posterior probability of H_0: Consonni and La Rocca example.

posterior probability of the null hypothesis. The posterior distributions are very little affected by the different priors.

From the 10,000 draws, the empirical posterior probability that the deviance difference is negative, that is, the null hypothesis is better supported than the alternative, is 0.035, 0.017, and 0.022 for the Haldane, Jeffreys, and uniform priors, respectively. These values are similar, but not close, to the *p*-value from the likelihood ratio test.

Converting deviances to likelihood ratios and posterior probabilities, the *median* posterior probabilities of the null hypothesis for the three priors are around 0.1, much smaller than the values from the Bayes factors, but the corresponding 95% central credible intervals are very skewed and diffuse: [0.078, 0.519], [0.078, 0.497], and [0.078, 0.484], respectively. The various Bayes factors give posterior probabilities of the null hypothesis between the 85th and the 90th percentile of the posterior distribution.

These wide credible intervals for $\Pr[H_1 \mid \mathbf{y}]$ give quite different conclusions from the point values from the various Bayes factors. As noted in Chapter 2, the integrated likelihood in the denominator of the Bayes factor is the prior *mean* of the likelihood, but the prior *variability* about the mean is not reported, so the user is left with the impression that this *is* the likelihood required, rather than just a one-point summary of its prior distribution. The wide credible intervals make clear that statements about the posterior probability of the null hypothesis made through the Bayes factor are both *biased* (as one-point summaries) and *misleadingly precise* in this example.

6.4   Multiple binomial samples

6.4.1   A social network table

Table 6.2 is taken from Breiger (1974) and is originally from Davis et al. (1941).

The table classifies 18 South Carolina women by 14 social events which they did or did not attend. The table entries are 1 for attendance and 0 for no attendance, and the marginal totals T are shown. This table is *extremely sparse*. A first, and basic, question of interest is whether the events "woman" and "event" are *independent*, that is, whether the profile of attendance probabilities across events is the same for all women. The data table shows a clear pattern of event participation moving from the left to the right side of the table as we move downward: we expect to reject the hypothesis of independence.

There is no experimental design: the number of events attended by each woman is a random variable, not fixed. We index the table by $i = 1, \ldots, r$ for rows and $j = 1, \ldots, c$ for columns.

6.4.2   Network model

For these data, and for other social network data, the cell entry is a *Bernoulli* variable, since attendance either occurs or does not. The table thus consists of

TABLE 6.2

Event Attendance

W	1	2	3	4	5	6	E 7	8	9	10	11	12	13	14	T
1	1	1	1	1	1	1	0	1	1	0	0	0	0	0	8
2	1	1	1	0	1	1	1	1	0	0	0	0	0	0	7
3	0	1	1	1	1	1	1	1	1	0	0	0	0	0	8
4	1	0	1	1	1	1	1	1	0	0	0	0	0	0	7
5	0	0	1	1	1	0	1	0	0	0	0	0	0	0	4
6	0	0	1	0	1	1	0	1	0	0	0	0	0	0	4
7	0	0	0	0	1	1	1	1	0	0	0	0	0	0	4
8	0	0	0	0	0	1	0	1	1	0	0	0	0	0	3
9	0	0	0	0	1	0	1	1	1	0	0	0	0	0	4
10	0	0	0	0	0	0	1	1	1	0	0	1	0	0	4
11	0	0	0	0	0	0	0	1	1	1	0	1	0	0	4
12	0	0	0	0	0	0	0	1	1	1	0	1	1	1	6
13	0	0	0	0	0	0	1	1	1	1	0	1	1	1	7
14	0	0	0	0	0	1	1	0	1	1	1	1	1	1	8
15	0	0	0	0	0	0	1	1	0	1	1	1	1	1	7
16	0	0	0	0	0	0	0	1	1	1	0	1	0	0	4
17	0	0	0	0	0	0	0	0	1	0	1	0	0	0	2
18	0	0	0	0	0	0	0	0	1	0	1	0	0	0	2
T	3	3	6	4	8	8	10	14	12	6	4	7	4	4	93

an $r \times c$ array of Bernoulli variables y_{ij}, or binomials with sample sizes $n_{ij} = 1$ and attendance probabilities p_{ij}, with $0 \le p_{ij} \le 1$ for all (i, j). Constructing the model for the table of Bernoullis is not straightforward because it is not obvious that these can be treated as independent – we may expect attendance events to be correlated within a woman. For the moment we assume independence. The general "saturated" model under the alternative hypothesis H_2 is then

$$\Pr[\{y_{ij}\} \mid \{p_{ij}\}] = \prod_i \prod_j p_{ij}^{y_{ij}} (1 - p_{ij})^{1-y_{ij}}.$$

Under the null hypothesis model of *additivity* of woman and event, we have a fixed-effect version of the *Rasch* model of *item response theory*:

$$H_1 : \phi_{ij} = \text{logit } p_{ij} = \alpha_j + \theta_i,$$

where α_j is the "event attractiveness" and θ_i is the "woman attendance propensity."

The Rasch model likelihood is awkward to simulate from. We simplify it by using the "Poisson trick" (Aitkin et al. 2008, pp. 288–9). We treat the cell Bernoullis as Poisson counts, with the Rasch model being the additive model in a Poisson two-factor log-linear model. So under H_2, $Y_{ij} \sim P(\mu_{ij})$ with μ_{ij} unrelated, and under H_1, $\mu_{ij} = \alpha_j \cdot \theta_i$.

We need to fix the parameter scales, since equivalently

$$\mu_{ij} = [\alpha_j \cdot a] \cdot [\theta_i \cdot (1/a)]$$

for any a. We fix $\sum_j \alpha_j = 1$. Then

$$\mu_{ij} = \alpha_j \cdot \theta_i$$

$$\sum_j \mu_{ij} = \mu_{i+} = \sum_j \alpha_j \cdot \theta_i$$

$$= \theta_i$$

$$\sum_i \mu_{ij} = \mu_{+j} = \alpha_j \cdot \sum_i \theta_i$$

$$= \alpha_j \mu_{++}$$

$$\alpha_j = \mu_{+j}/\mu_{++}$$

$$\alpha_j \cdot \theta_i = \mu_{i+}\mu_{+j}/\mu_{++}$$

The likelihoods and log-likelihoods are (omitting the constant term $\prod y_{ij}! = 1$)

$$L_2 = \prod_i \prod_j \left[\exp(-\mu_{ij})\mu_{ij}^{y_{ij}} \right]$$

$$\log L_2 = -\mu_{++} + \sum_i \sum_j y_{ij} \log \mu_{ij}$$

$$L_1 = \prod_i \prod_j \left[\exp(-\alpha_j \theta_i)(\alpha_j \theta_i)^{y_{ij}} \right]$$

$$\log L_1 = -\sum_i \theta_i \cdot \sum_j \alpha_j + y_{+j} \log \alpha_j + y_{i+} \log \theta_i$$

$$= -\mu_{++} + y_{+j} \log \mu_{+j} + y_{i+} \log \mu_{i+} - y_{++} \log \mu_{++}.$$

6.4.3   Frequentist analysis

For the frequentist analysis, the model can be fitted by maximum likelihood with either the binomial or the Poisson model. For the Poisson model, the frequentist deviance for the saturated model is 186.00, and for the additive model is 335.09. The LRTS is 149.09 with 221 "degrees of freedom"; the p-value from χ_{221}^2 is 0.9999!

The χ_{221}^2 distribution assumption is clearly *invalid*: the table is *essentially sparse*, since the "Poisson count" can only be 0 or 1. More women and more events do not change this. We have no frequentist way of interpreting the LRTS, except by evaluating all permutations of the 1s and 0s conditional on the fixed margins and determining the probability of a "more extreme" table under the null hypothesis (the generalized Fisher "exact" test). This permutation test does not use the full likelihood.

6.4.4   Choice of alternative model prior

The choice of conjugate gamma prior is critical because the sample "counts" y_{ij} in each cell are 0 or 1. The uniform or Jeffreys priors are highly informative (relative to the 0/1 count). We use the improper prior $\pi(\mu) = \mu^{a-1}$ with

$a = 1/rc$ for each cell. The cell posterior is then

$$\pi(\mu \mid y) = \exp(-\mu)\mu^{y+a-1}/\Gamma(y+a),$$

which is proper so long as $y + a > 0$. The total prior sample size across all cells is 1 (compared with the 93 attendance 1s).

6.4.5 Simulations

We make M random draws $\mu_{ij}^{[m]}$ from the gamma posterior with indices $y_{ij}+a$, and substitute in the deviance $D_2 = -2\log L_2$ to give M random draws $D_2^{[m]}$ from D_2. At each draw m we calculate the marginal sums

$$\mu_{i+}^{[m]} = \sum_j \mu_{ij}^{[m]}, \ \mu_{+j}^{[m]} = \sum_i \mu_{ij}^{[m]}, \ \mu_{++}^{[m]} = \sum_i \mu_{i+}^{[m]}$$

and substitute them in $D_1 = -2\log L_1$ to give the corresponding M draws $D_1^{[m]}$. The M deviance difference draws are then given by $D_{12}^{[m]} = D_1^{[m]} - D_2^{[m]}$. The deviance distributions for the null (dotted curve) and alternative (solid curve) models are shown in Figure 6.3.

The frequentist deviances are the circles at the left-hand end of each distribution. That for the saturated model is far from the posterior distribution of deviances from this model. For the additive model the frequentist deviance is much closer to the end of the posterior distribution. The deviance difference distribution is shown in Figure 6.4.

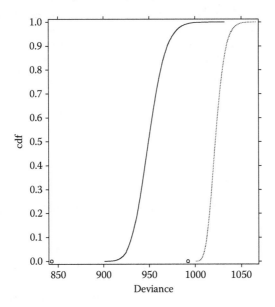

FIGURE 6.3
Deviances: Davis example.

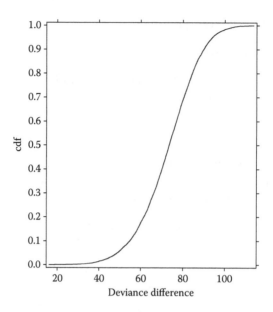

FIGURE 6.4
Deviance difference: Davis example.

The entire difference distribution is well above 0 (the smallest value is 15.9), but it is also well *below* the LRTS value of 143. The 95% credible interval is [43.57, 97.83]. We can very firmly reject the hypothesis of additivity.

The purpose of the original study, and of later analyses of these data, was to investigate whether *social networks* existed in this population. For the data, can we determine whether a subset of women attended a subset of the same events, suggesting a network of relationships among such women? The rejection of the null hypothesis of additivity is only a (necessary) first step, and does not lead to a conclusion on this question, but this can be assessed by fitting *latent class models* which define such a classification of women into (latent) groups.

6.5 Two-way tables for categorical responses – no fixed margins

We discuss two different types of two-way table: those with only the total sample size fixed – no marginal totals are fixed – and those with one margin fixed by a sample design. The case of both margins fixed is very rare, and we do not deal with it here.

In the frequentist analysis, these two types of tables are treated in the same way, but this is not so in the Bayesian approach, because the different

TABLE 6.3

Babies Surviving or Dying Under
CMT and ECMO

Treat	Response Survived	Died	Total
CMT	0	1	1
ECMO	11	0	11
TOTAL	11	1	12

multinomial models involved have different posterior behavior. We consider first the case of a single multinomial defined by two cross-classifying variables.

6.5.1 The ECMO study

The data come from a randomized clinical trial of ECMO (extra-corporeal membrane oxygenation) for the treatment of newborn babies with respiratory distress, compared to the then-current best medical treatment (CMT). The study was reported in Bartlett et al. (1985), and statistical and ethical issues in this trial, and in a subsequent trial, were discussed in Ware (1989) and Begg (1990). The trial used adaptive randomization of babies to the treatments, with a success (survival) under a treatment increasing the probability of randomization of the next baby to that treatment. This was intended to minimize the number of babies randomized to the less effective treatment. The trial data are given in Table 6.3.

The frequentist analysis was complicated by the adaptive randomization, and by the stopping rule of the study, which was to stop when the difference in the number of babies surviving between the treatments was 9. As is clear, this rule was not followed: at the specified stopping time, the study continued with two babies assigned nonrandomly (i.e., with probability 1) to the ECMO condition. Both survived.

Many *p*-values were given for this table (Ware 1989; Begg 1990) by different frequentist arguments: from 0.001 to 0.62.

6.5.2 Bayes analysis

The table *looks* as though a two-binomial model is appropriate, but the adaptive randomization makes the marginal treatment sample sizes *random*, not fixed, so we first use a *multinomial* model.

We index the (general two-way) table by $i = 1, \ldots, r$ for rows and $j = 1, \ldots, c$ for columns. The probability that treatment i gives outcome j is p_{ij}, with $\sum_i \sum_j p_{ij} = 1$; n_{ij} is the number treated under i giving outcome j.

The null hypothesis of independence is $H_1 : p_{ij} = p_{i+} \cdot p_{+j}$, where

$$p_{i+} = \sum_j p_{ij}, \; p_{+j} = \sum_i p_{ij}$$

6.5.3   Multinomial likelihoods

The likelihoods under the alternative and null hypotheses are

$$L_2 = \prod_i \prod_j p_{ij}^{n_{ij}}$$

$$L_1 = \prod_i \prod_j [p_{i+} \cdot p_{+j}]^{n_{ij}}$$

$$= \prod_i p_{i+}^{n_{i+}} \cdot \prod_j p_{+j}^{n_{+j}}$$

The LRT maximizes the likelihoods under each hypothesis: the LRTS is 6.88, with a p-value of 0.0087 from χ_1^2. The Pearson X^2 is 12.0, with a p-value of 0.0006 from χ_1^2. In this extreme case, the Pearson X^2 value is equal to the total sample size, which seems uninformative about the difference in response probabilities. A separate question is whether the asymptotic distribution is valid, even assuming the correctness of the usual multinomial analysis.

One of the many questions argued over in the discussions of this paper is whether this is the correct frequentist analysis. Frequentist analysts had to decide whether and how the adaptive randomization should be allowed for: by conditioning on the assignment sequence, or not, or by excluding the two nonrandomly assigned babies, or not, and if not, what other conditioning, if any, should be used. Different conditioning approaches lead to different conditional p-values. In particular, conditioning on the treatment assignment *and* the attained sample sizes reduces the problem to the usual two-sample binomial comparison.

These arguments illustrate a *critical advantage of Bayes analysis – the stopping rule is irrelevant* as long as it is *uninformative* about the response probabilities. That is the case here: the assignment probability for each baby to each treatment was based on the numbers surviving amongst those already treated, so is a known number at each step. So the treatment assignment process with its varying randomization probabilities and the resulting random sample sizes are *ignorable* in the technical missing data sense – they can be omitted from the likelihood. (If they are *retained*, the assignment probabilities *cancel* from the likelihood ratio, and so play no role in inference, as discussed in earlier chapters.)

This is an important issue for the Bayes analysis as well: we do not need the additional complexity of modeling the assignment process in the full multinomial model. We compare below the multinomial results with the two-binomial analysis from Aitkin et al. (2005).

6.5.4   Dirichlet prior

For the Bayesian multinomial analysis, we use a conjugate Dirichlet prior:

$$\pi(\{p_{ij}\} \mid \{a_{ij}\}) = \frac{\Gamma(a_{++})}{\prod_i \prod_j \Gamma(a_{ij})} \prod_i \prod_j p_{ij}^{a_{ij}-1}.$$

The noninformative (Haldane) prior has $a_{ij} = 0 \,\forall i, j$. A difficulty with this prior is that it restricts the posterior to the observed (nonzero) support, and in effect treats the zero counts as *structural* – an essential feature of the model – rather than as *sampling* zeros, a consequence of the random assignment process. The posterior simulations then come *only* from the cells with nonzero counts, a severe limitation in this extreme table.

For a weakly informative prior, we take the prior indices $a_{ij} = a = 1/rc = 0.08$ for all (i, j): this corresponds to a "prior sample size" of $a_{++} = 0.32$, compared to the sample weight of 12.

6.5.5 Simulations

We make M random draws $p_{ij}^{[m]}$ from the Dirichlet posterior with indices $n_{ij} + a$, and substitute in the alternative model deviance $D_2 = -2 \log L_2$ to give M random draws $D_2^{[m]}$ from D_2. At each draw m we calculate the marginal probabilities

$$p_{i+}^{[m]} = \sum_j p_{ij}^{[m]}, \quad p_{+j}^{[m]} = \sum_i p_{ij}^{[m]},$$

and substitute them in $D_1 = -2 \log L_1$ to give the corresponding M draws $D_1^{[m]}$. The M deviance difference draws are then given by $D_{12}^{[m]} = D_1^{[m]} - D_2^{[m]}$. The deviances are shown in Figure 6.5, and the deviance differences are shown

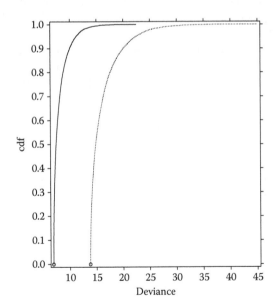

FIGURE 6.5
Deviances: ECMO example.

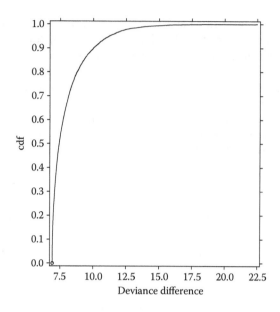

FIGURE 6.6
Deviance difference: ECMO example.

in Figure 6.6, for the weakly informative Dirichlet prior and $M = 10,000$ draws.

All 10,000 values of the deviance difference are positive, and nearly 97% are larger than the LRTS. The null hypothesis is *firmly rejected*. But could this be the consequence of the prior?

We carry out a small sensitivity study to assess the effect of the prior. We repeat the same analysis with the common prior parameter set to $a = 0.25$ – a much larger prior sample size relative to the data sample size – and to zero, giving the Haldane prior. Table 6.4 gives for the three priors the proportion p of negative deviance differences (supporting the null hypothesis), together with the smallest and the 2.5, 50, and 97.5 percentiles of the empirical distribution of deviance differences.

For the prior parameters $a = 0.25$, the proportion of negative differences is 0.0028. All the 95% credible intervals exclude zero deviance difference, so it is clear that ECMO was better.

TABLE 6.4

Effect of Priors

a	0	0.08	0.25
p	0.0000	0.0000	0.0028
$DD_{(1)}$	6.88	2.93	-10.42
$DD_{(250)}$	6.88	6.83	3.27
$DD_{(5000)}$	7.0	7.38	6.84
$DD_{(9750)}$	12.44	12.46	10.38

To determine the *size* of the difference in response probabilities we need to adapt the simulation results to these probabilities. We have the draws from the posterior distribution of the multinomial parameters p_{11}, p_{12}, p_{21}, and p_{22}. Defining i as the treatment index with $i = 1$ for CMT and 2 for ECMO, and j as the response index with $j = 1$ for survival, the difference between the *conditional probabilities* of survival θ_1 given ECMO and θ_2 given CMT is

$$\delta = \theta_2 - \theta_1$$
$$= \frac{p_{21}}{p_{2+}} - \frac{p_{11}}{p_{1+}}$$
$$= \frac{p_{21}p_{12} - p_{11}p_{22}}{p_{1+}p_{2+}}.$$

The Dirichlet prior and posterior transform correspondingly:

$$\pi(\theta_1) = \theta_1^{a_{11}-1}(1-\theta_1)^{a_{12}-1}/B(a_{11}, a_{12})$$
$$\pi(\theta_1 \mid \mathbf{y}) = \theta_1^{a_{11}+n_{11}-1}(1-\theta_1)^{a_{12}+n_{12}-1}/B(a_{11}+n_{11}, a_{12}+n_{12})$$
$$\pi(\theta_2) = \theta_2^{a_{21}-1}(1-\theta_2)^{a_{22}-1}/B(a_{21}, a_{22})$$
$$\pi(\theta_2 \mid \mathbf{y}) = \theta_2^{a_{21}+n_{21}-1}(1-\theta_1)^{a_{22}+n_{22}-1}B(a_{21}+n_{21}, a_{22}+n_{22})$$

with θ_1 independent of θ_2. These are just the standard beta results for the binomial distributions.

The effect of the noninformative Haldane prior with $a_{ij} = 0$ is to give *degenerate* posteriors for both θ_1 and θ_2, because for θ_1, $n_{11} = 0$ and $n_{12} = 1$, giving a Beta(0,1) posterior, and for θ_2 with $n_{21} = 11$ and $n_{22} = 0$ the posterior is Beta(11,0). *All* the simulations from θ_1 will be zero, and all those from θ_2 will be 1!

We show in Table 6.5 the effect of increasing the prior weight a from 0 to 1 – from the Haldane through the Jeffreys to the uniform prior – on the posterior probability of a negative difference, the median difference, and the 95% central credible interval for the true difference in survival probabilities (ECMO-CMT).

TABLE 6.5

Prior Sensitivity

a	$\Pr[\delta < 0]$	Median	95% Interval
0	0	1.0	[1.000, 1.000]
0.1	0.0007	0.992	[0.258, 1.000]
0.2	0.0023	0.946	[0.165, 1.000]
0.3	0.0038	0.894	[0.122, 1.000]
0.4	0.0047	0.839	[0.104, 0.998]
0.5	0.0066	0.783	[0.097, 0.993]
0.6	0.0086	0.741	[0.083, 0.988]
0.7	0.0076	0.712	[0.075, 0.981]
0.8	0.0096	0.676	[0.073, 0.969]
0.9	0.0110	0.650	[0.062, 0.958]
1.0	0.0115	0.620	[0.056, 0.948]

The effect of increasing prior weight is clear: a progressive increase in the posterior probability of a negative difference (that CMT is better than ECMO), though this probability does not rise above 0.012 for the uniform prior, and a progressive increase in the uncertainty about the true difference. The two-binomial analysis in Aitkin et al. (2005) comes to the same conclusion: with the uniform priors on θ_1 and θ_2 used there, the empirical probability that CMT is better than ECMO is 0.011 (with $M = 10,000$).

With such a small sample size under the CMT treatment we cannot give any precise statement of the difference in the response probabilities, except that it is positive. A larger study was needed for a more precise statement about this difference, or else a strongly informative prior for the CMT treatment response probability. This illustrates the importance of a study design which allows convincing conclusions to be drawn: some of the discussants of the two papers on the study commented on its ethics.

6.6 Two-way tables for categorical responses – one fixed margin

We now consider the case when one table margin is fixed by an experimental design. Table 6.6 and the description of the study are taken from the StatXact Web site (StatXact, Example 2).

> Researchers were provided four types of rewards for successful comple-
> tion of a block building task. The rewards may be categorized as achieve-
> ment oriented, financial, socially reinforcing, and neutral. Five subjects
> were assigned at random to each reward category. The subjects were
> instructed in advance about the respective reward structures, and then
> assigned to the block building task. Each subject was interrupted in a
> standard way while carrying out the task. The object of the experiment
> was to determine if the subject's reaction to task disruption was related
> to the type of reward.

Since the number of subjects assigned to each reward condition is fixed, the models under the null and alternative hypotheses are different from

TABLE 6.6

Reaction To Task Disruption

Type of Reward	Start over	Abandon Task	Modify Task	Totals
Achievement Oriented	5	0	0	5
Financial	2	1	2	5
Socially Reinforcing	2	0	3	5
Neutral	0	1	4	5
Totals	9	2	9	20

the multinomial. We use i to index rows (rewards), j to index reactions (responses), and n_{ij} is the number of category j responses in reward condition i. We now use p_{ij} to denote the probability of reaction category j under reward condition i, with $\sum_j p_{ij} = 1$. Under the alternative hypothesis H_1, these probabilities are general, and unrelated across reward conditions i. The likelihood under the alternative model is then

$$L_2 = \prod_i \prod_j p_{ij}^{n_{ij}}.$$

This *appears* to be the same likelihood as in the previous case, but now we have r sets of multinomial probabilities p_{ij} with $\sum_j p_{ij} = 1$ rather than a *single* multinomial. The frequentist deviance for this model is

$$
\begin{aligned}
FD_2 &= -2 \sum_i \sum_j n_{ij} \log \hat{p}_{ij} \\
&= -2 \sum_i \sum_j n_{ij} \log[n_{ij}/n_{i+}] \\
&= -2[\sum_i \sum_j n_{ij} \log n_{ij} - \sum_i n_{i+} \log n_{i+}] \\
&= 22.28.
\end{aligned}
$$

Under the null hypothesis H_1 of *homogeneity* (across reward conditions), the category response probabilities are the same: $p_{ij} = q_j$ for all i. The frequentist deviance for this model is

$$
\begin{aligned}
FD_1 &= -2 \sum_i \sum_j n_{ij} \log \hat{q}_j \\
&= -2 \sum_i \sum_j n_{ij} \log[n_{+j}/n_{++}] \\
&= -2[\sum_j n_{+j} \log n_{+j} - n_{++} \log n_{++}] \\
&= 37.96.
\end{aligned}
$$

The likelihood ratio test statistic is 15.67; if this could be compared with χ_6^2, its p-value would be 0.0156. The StatXact site gives the Pearson X^2 test statistic of 11.56; if *this* could be compared with χ_6^2, its p-value would be 0.0725.

For the Bayes analysis, the likelihood under the nested null model has to be expressed in terms of the multinomial probabilities under the alternative model. We define

$$q_j = \frac{\sum_i n_{i+} p_{ij}}{\sum_i n_{i+}},$$

the population version of the MLE of q_j. The likelihood under the null model is then

$$L_1 = \prod_j q_j^{n_{+j}}.$$

Computation of the likelihood ratio distribution follows as in the previous case. Because of the small sample sizes of 5 in each reward condition, we use the minimally informative Dirichlet prior with $a_{ij} = a = 0.1/3$ for all i and j, giving a prior sample size of 0.1 in each reward condition. We make M random draws $p_{ij}^{[m]}$ from the Dirichlet posterior with indices $n_{ij} + a$, and substitute them in the deviance $D_2 = -2 \log L_2$ to give M random draws $D_2^{[m]}$ from D_2. At each draw m we calculate the marginal probabilities

$$q_j^{[m]} = \frac{\sum_i n_{i+} p_{ij}^{[m]}}{\sum_i n_{i+}},$$

and substitute them in $D_1 = -2 \log L_1$ to give the corresponding M draws $D_1^{[m]}$; the M deviance difference draws are then given by $D_{12}^{[m]} = D_1^{[m]} - D_2^{[m]}$.

The deviance distributions for the homogeneity (dotted curve) and saturated (solid curve) models are shown in Figure 6.7.

The frequentist deviances are shown as circles at the left-hand end of each distribution; both frequentist values are very close to the minimum sampled deviances. However, the posterior distributions do not agree well with the

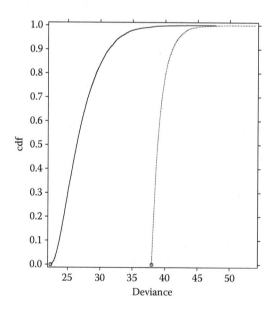

FIGURE 6.7
Deviances: StatXact example.

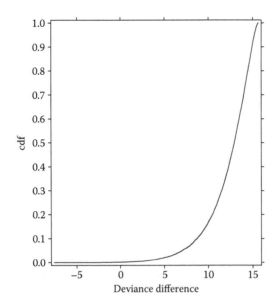

FIGURE 6.8
Deviance difference: StatXact example.

asymptotic shifted χ_2^2 (homogeneity)) and χ_8^2 (saturated) distributions (not shown): the cell zeros do not provide the corresponding posteriors with an internal mode. The deviance difference distribution is shown in Figure 6.8.

Of the 10,000 deviance differences, 15 are negative, so the empirical posterior probability that the homogeneity model has higher likelihood than the saturated model is 0.0015. The 95% central credible interval for the true deviance difference is [5.47, 15.36]; the corresponding interval for the posterior probability of the null hypothesis, given equal prior probabilities, is [0.0005, 0.061]. The observed LRTS is beyond the upper 97.5% point of the posterior distribution, and so greatly overstates the size of the true likelihood ratio. Nevertheless, the posterior evidence against the homogeneity hypothesis is very strong.

The StatXact Web site gives a *p*-value of 0.0398 from the "exact" (conditional) test by enumerating the sample configurations which would give "more extreme" values of the LRTS than that observed in the data. This is much weaker evidence than that provided by the Bayes analysis.

It might be thought that the minimally informative Dirichlet prior might be responsible for this result. We give the corresponding results below for the noninformative Haldane prior with parameters $a_{ij} = 0$ for all (i, j); here the prior sample size is 0 instead of 0.1. The graphs of the deviance distributions (not shown) are indistinguishable from those shown in Figure 6.7. There were 17 negative deviance differences, and the 95% credible intervals for the true deviance difference is [5.65, 15.43], giving the corresponding interval for the posterior probability of the null hypothesis, given equal prior probabilities, as [0.0004, 0.056]. The conclusions hardly differ from those for the minimally informative prior.

6.7 Multinomial "nonparametric" analysis

The multinomial/Dirichlet approach can also be used, as we described in detail in Chapter 4, for "continuous" variables. We give a simple example, as an alternative to the Wilcoxon–Mann–Whitney two-sample test.

The data in Table 6.7 come from an example in Hodges et al. (1975, p. 206). They are weight gains in pounds, in increasing order, of two samples of Holstein calves, randomly assigned to either a control diet or the same diet with a supplement. The question is whether the supplement is effective in increasing weight, above that from the control diet.

The two-sample t test gives a (one-sided) p-value of 0.058 (for the alternative hypothesis $\mu_S > \mu_C$), while the Wilcoxon–Mann–Whitney test gives a one-sided p-value of 0.046. (The p-values for the usual general alternative $\mu_S \neq \mu_C$ are twice the one-sided p-values.)

For Bayes inference, the mean difference in weight gain is defined by

$$\delta = \mu_C - \mu_S = \sum_J p_{CJ}\, y_J - \sum_J p_{SJ}\, y_J .$$

We use a slightly informative Dirichlet prior on the observed support across both samples, with $a_J = 0.02$ for observed values and 0 for unobserved values. We show in Figure 6.9 the posterior distributions for the control and supplement means, and in Figure 6.10 the posterior distribution of the control-supplement mean difference, from $M = 10,000$ draws.

The posterior median of the difference is -17.1 pounds, and the 95% central credible interval is $[-37.0, 0.5]$ pounds. Of the 10,000 draws, 297 are negative, giving an empirical posterior probability of 0.970 (with standard error 0.0017) that the mean difference is negative.

We show also in Figure 6.11 the posterior distributions of the deviances for the null model of a common multinomial distribution in the two groups and for the alternative model of different multinomials, and in Figure 6.12 the deviance difference between the null and alternative models.

TABLE 6.7

Weight Gains in Pounds

Control	Supplement
53	66
57	73
62	77
68	81
71	90
74	114
84	132
109	

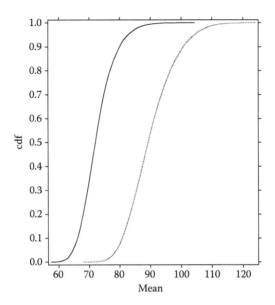

FIGURE 6.9
Mean posteriors: diet example.

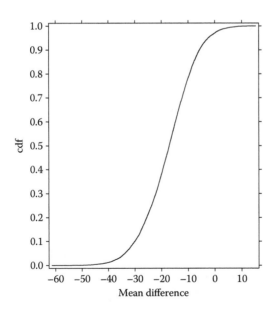

FIGURE 6.10
Mean difference posterior: diet example.

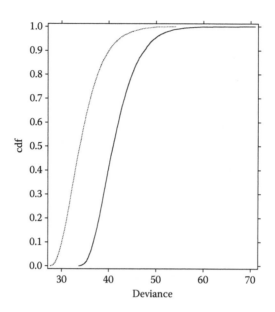

FIGURE 6.11
Deviance posteriors: diet example.

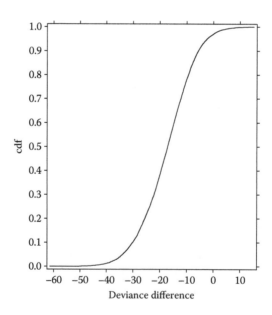

FIGURE 6.12
Deviance difference posterior: diet example.

The median deviance difference is 7.03, and the 95% credible interval is $[-5.04, 19.16]$. Of the 10,000 draws, 1089 are negative, giving a Bayesian p-value of 0.109 for the null hypothesis of a common multinomial distribution. This does not correspond to the posterior probability of a positive mean difference, as the hypotheses being compared are quite different: means and multinomial distributions. For inference about the mean difference, the Bayesian bootstrap analysis gives slightly stronger evidence that the diet supplement is better than the control.

7

Goodness of Fit and Model Diagnostics

This chapter is based partly on the unpublished paper (Aitkin 2007): "Bayesian goodness of fit assessment through posterior deviances." We are concerned mainly with Bayesian goodness of fit assessment and diagnostics for the *probability model assumption*, but first consider frequentist methods.

7.1   Frequentist model diagnostics

These are discussed in many books; Aitkin et al. (2009) discuss the exponential family distributions in some detail. Distributions with sufficient statistics for the model parameters provide useful checks on some aspects of the model through *ancillary statistics*. These are functions of the data whose distributions do not depend on the model parameters, so contribute nothing to the likelihood. By conditioning the observed data on the sufficient statistics for the parameters, we eliminate these from the conditional distribution and obtain a *completely specified* conditional distribution.

For example, for a binomial sample of size $n = 6$ giving the sequence SFFSSS, with four successes and two failures with success probability θ, the likelihood function is the probability of the observed sequence:

$$L(\theta \mid \text{SFFSSS}) = \Pr[\text{SFFSSS} \mid \theta]$$
$$= \theta \cdot (1 - \theta) \cdot (1 - \theta) \cdot \theta \cdot \theta \cdot \theta$$
$$= \theta^4 (1 - \theta)^2.$$

The sufficient statistic for θ is r, the number of successes, which has the binomial $b(6, \theta)$ distribution. The *conditional* distribution of the observed sequence *given r* is

$$\Pr[\text{SFFSSS} \mid r, \theta] = \Pr[\text{SFFSSS} \mid \theta] / \Pr[r \mid \theta]$$
$$= 1 / \binom{6}{4}.$$

It is not immediately obvious what this means, or how it could be used. Since every possible sequence would give the same conditional probability, the conditional distribution is relevant to *arrangements* of the successes and failures within the sequence. The sequence FFSSSS might lead us to suspect (depending on the context) a systematic factor at work as a result of which the success probability changes during the sequence. It is clear that "suspicious" (nonrandom) patterns of Ss and Fs would be those with an unusual number of *runs* of the same outcome, either a *small* number (as in FFSSSS), or a *large* number (as in SFSFSSS), so an assessment of the probability of the observed numbers of runs of different length, under a random sequence "model," would be one way of investigating a suspicious pattern. The data themselves *cannot lead to this conclusion,* but can suggest that it might be worth investigating.

The Poisson distribution has a different conditional property. For a random sample $y_1, \ldots, y_n$ from $P(\mu)$, the probability of the observed sample is

$$\Pr[y_1, \ldots, y_n \mid \mu] = \prod_{i=1}^{n} e^{-\mu} \mu^{y_i} / y_i!$$

$$= e^{-n\mu} \mu^T / \prod_i y_i!,$$

where $T = \sum y_i$. The sufficient statistic T has a Poisson $P(n\mu)$ distribution, so the conditional distribution of $y_1, \ldots, y_n$ given T is

$$\Pr[y_1, \ldots, y_n \mid T, \mu] = \Pr[y_1, \ldots, y_n \mid \mu] / \Pr[T \mid \mu]$$

$$= \frac{T!}{\prod_i y_i!} \left(\frac{1}{n}\right)^T$$

$$= \frac{T!}{\prod_i y_i!} \prod_i \left(\frac{1}{n}\right)^{y_i},$$

a multinomial distribution with n category probabilities $1/n$ and category event counts y_i. It is clear that a large outlying value of y_i will give a small multinomial probability, but it is unclear how the Poisson model assumption is to be assessed, in terms of "what might be expected" from this multinomial distribution.

In continuous location/scale models like the normal and extreme value, the *residuals* have a similar property. For the normal model, the sufficient statistics are $\bar{y}$ and RSS. The (sampling) distributions of these statistics are $N(\mu, \sigma^2/n)$ and $\sigma^2 \chi_v^2$, with $v = n - 1$, and the conditional distribution of the sample values $y_1, \ldots, y_n$ given $\bar{y}$ and RSS is

$$f(y_1, \ldots, y_n \mid \bar{y}, RSS) = \frac{\Gamma(v/2)}{\pi^{v/2}} \cdot \frac{1}{RSS^{v/2-1}}.$$

Some insight into this can be obtained from the Studentized residuals $(y_i - \bar{y})/s$:

$$y_i - \bar{y} \sim N(0, (n-1)\sigma^2/n)$$

$$\sqrt{n}(y_i - \bar{y})/[\sqrt{n-1}]\sigma \sim N(0, 1)$$

$$\sqrt{n}(y_i - \bar{y})/[\sqrt{n-1}s] \sim t_v.$$

However, the residuals are correlated, with common correlation $-1/(n-1)$, and have a (scaled) multivariate t-distribution. *Residual probability plotting* is widely used to check the probability model assumption: eliminating the unknown parameters gives a known distribution for the residuals which can then be assessed through the empirical cdf of the residuals. The common negative correlation of the residuals is usually ignored, and a normal plot of the Studentized or raw residuals is commonly used; by drawing repeated samples of the given sample size from the known distribution, we can construct *simulation envelopes* from the cdfs of the samples, within which the actual sample empirical cdf should fall with high probability. Atkinson (1985) gave a discussion of this approach.

Assessing the goodness of fit of a discrete distribution like the Poisson requires a different approach. In the frequentist framework the Pearson X^2 test is the most commonly used method; here the category probabilities implied by the assumed model are compared with the observed category proportions – the maximum likelihood estimates under a general alternative multinomial model.

This method has well-known limitations: "continuous" data have to be grouped arbitrarily into categories, and category counts cannot be too small for the validity of the asymptotic χ^2 distribution. Small counts naturally occur in the tails of the distribution, where departures from the model may be most important, so the need to group categories in the extreme tails leads to reduced power of the test to detect such tail departures.

There are many other frequentist tests for goodness of fit, some formulated for specific classes of departures from the assumed parametric model. These were reviewed in d'Agostino and Stephens (1986); a recent brief survey is given in Conigliani et al. (2000), together with a detailed survey of Bayesian methods for this problem.

7.2 Bayesian model diagnostics

Bayesian procedures for assessing parametric model assumptions are much less developed. There are two general approaches.

The first computes a Bayes factor for the null parametric model compared to the general alternative multinomial model. The Bayes factor has well-known difficulties with diffuse priors (described, for example, by Conigliani et al.

2000), and elicitation of personal priors for the model parameters is difficult. The fractional Bayes factor (O'Hagan 1995) was examined as an alternative Bayesian method for model assessment (Conigliani et al. 2000), but was
found to give conflicting assessments compared to well-established frequentist methods. Reformulating it through a hierarchical model gave reasonable
results for discrete data, but did not provide an analysis for continuous data.
An extension to the "generalized fractional Bayes factor" by Spezzaferri et al.
(2007) improved the discrete data performance of the fractional Bayes factor,
but did not deal with continuous data.

The second general approach uses the posterior predictive *p*-value – the
ppp – proposed by Rubin (1984) and Gelman et al. (1996). This computes a
tail area probability, of an appropriate measure of discrepancy between the
data and the model, from the posterior predictive distribution. This method
has become popular and is now widely used. However, *calibration* of the
ppp is unclear, and it has been found to suffer severe *mis*-calibration (Hjort
et al. 2006), in the sense that in large samples it tends to be 0.5 in probability,
regardless of the actual data.

In the remaining sections of this chapter, we first consider the *ppp* and point
out difficulties with the posterior predictive distribution which closely parallel those of Bayes factors. We then follow closely the alternative model approach of Conigliani et al. (2000) with a multinomial distribution and Dirichlet
prior for the alternative model, but compare the models through their posterior deviance distributions rather than through the fractional Bayes factor,
following Aitkin (2007). This approach works equally well for continuous and
for discrete data, and we apply it to several examples given in Conigliani et al.
The approach also works well for comparing general nonnested competing
models, as we demonstrate with a large-scale simulation study of four models.

The computational method requires no more than standard Monte Carlo
simulations from the Dirichlet and other standard posterior distributions; we
use diffuse priors for the parameters of the distributions being compared.

7.3   The posterior predictive distribution

We are given data $\mathbf{y} = (y_1, \ldots, y_n)$ from a model $f(y \mid \theta)$, with a prior distribution $\pi(\theta)$, and wish to make a predictive statement about *new* data from the
same model. We consider only the case of a single new observation denoted
by y_0. How should we use the given data for this purpose?

From the data we have the likelihood $L(\theta)$ and the posterior distribution
$\pi(\theta \mid \mathbf{y})$ of θ. The standard Bayesian method for inference about y_0 is to
integrate out θ from the probability of the new data with respect to its posterior
distribution, to give the *posterior predictive distribution* of y_0:

$$f^*(y_0 \mid \mathbf{y}) = \int f(y_0 \mid \theta)\pi(\theta \mid \mathbf{y})d\theta.$$

We use f^* to denote the integrated probability mass or density function of y_0 because this will in general not be of the same form as $f(y \mid \theta)$.

We illustrate with a simple binomial example. We observe r successes in n Bernoulli trials with probability θ; the prior distribution of θ is uniform. What can we say about the probability of success at the next trial?

The posterior distribution of θ is beta$(r + 1, n - r + 1)$, and integrating out θ with respect to the posterior distribution gives

$$Pr[y_0 = 1] = \int Pr[y_0 = 1 \mid \theta]\pi(\theta \mid y)d\theta$$

$$= \int \theta^{r+1}(1 - \theta)^{n-r} d\theta / B(r + 1, n - r + 1)$$

$$= B(r + 2, n - r + 1)/B(r + 1, n - r + 1)$$

$$= (r + 1)/(n + 2).$$

We have achieved a curious result here: the probability of a success before the data were observed was the unknown parameter θ with prior $\pi(\theta)$, and after n trials this is updated to the beta posterior, but the probability of a success at the next trial has become a *known* probability! By observing the data – *any data* – we have *completely removed* the parametric uncertainty in any probability statement about new data. How is this possible?

It is not. To see this, consider a different inferential statement about y_0. From the posterior distribution of θ, we can construct a 95% central credible interval for θ:

$$Pr[\theta_{0.025} < \theta < \theta_{0.975} \mid \mathbf{y}] = 0.95.$$

Since the probability of success at *any* trial is θ, it follows that

$$Pr[\theta_{0.025} < Pr[y_0 = 1 \mid \mathbf{y}] < \theta_{0.975} \mid \mathbf{y}] = 0.95,$$

that is, the 95% credible interval for θ is also a 95% credible interval for *the probability of success at the next trial*; in fact *the posterior distribution of this probability is exactly the same as that of θ itself.*

If this is so, what is the status of the posterior predictive distribution of y_0? That is simply stated: it is the *posterior mean* of the distribution of y_0:

$$E_{\theta|y}\{Pr[y_0 = 1 \mid \theta]\} = (r + 1)/(n + 2).$$

What about the posterior *variance* of this distribution? This is

$$Var_{\theta|y}\{Pr[y_0 = 1 \mid \theta]\} = \int [\theta - (r + 1)/(n + 1)]^2 \pi(\theta \mid y)d\theta$$

$$= [(r + 1)(n - r + 1)]/[(n + 3)(n + 2)^2].$$

Suppose we observe $r = 10$ successes in $n = 10$ trials. What can we say about the probability of success at the 11th trial? The predictive distribution – the mean of the posterior distribution of y_{11} – is $\Pr(y_{11} = 1 \mid \mathbf{y}) = 11/12 = 0.917$. But the variance of this posterior distribution is $11/[13 \times 12^2] = 0.005876$, with standard deviation 0.077. The distribution is heavily skewed, so it is more useful to quote the 95% central credible interval: [0.715, 0.998]. This *correctly* represents our uncertainty about the probability of success at the next trial, which the predictive distribution does not.

Now consider the general case: we observe r successes in n trials, and want to make a statement about the probability of s successes in the next k trials. From the beta $(r+1, n-r+1)$ posterior for θ (assuming here a uniform prior) we make M draws $\theta^{[m]}$ and substitute them into the binomial probability $b(s \mid k, \theta)$ to give M draws $b^{[m]}(s \mid k, \mathbf{y}) = b(s \mid k, \theta^{[m]})$. These define the empirical posterior distribution of $b(s \mid k, \theta)$ given r and n.

For example, we observe $r = 1$ success in $n = 10$ trials. What can we say about the probability of at most 2 successes in the next 20 trials?

The predictive probabilities are easily calculated. The predictive probability of s successes in the next k trials is

$$p(s \mid k, r, n) = \binom{k}{s} \int \theta^s (1-\theta)^{k-s} \cdot \theta^1 (1-\theta)^9 / B(2, 10)$$

$$= \binom{k}{s} B(s+2, k-s+10)/B(2, 10)$$

which gives values of 0.118 for $s = 0$, 0.163 for $s = 1$, and 0.166 for $s = 2$. Thus the predictive probability of at most two successes in the next 20 trials is 0.447.

We make $M = 10{,}000$ draws from the beta $(2,10)$ posterior for θ and substitute them into the lower-tail probability of at most two successes in the binomial distribution $b(20, \theta)$. The posterior distribution of θ is shown in Figure 7.1, and that of the probability of at most two successes in shown in Figure 7.2.

The posterior distribution of the probability of at most two successes is remarkably diffuse, with median 0.413 but 95% credible interval [0.0023, 0.990] – almost the whole [0,1] range! The reason is clear from the posterior distribution of θ: this is also quite diffuse, with median 0.148 and 95% credible interval [0.023, 0.417]. The extremes of the interval for θ define those for the other interval, and it is clear that for the combination of a small observed sample and a larger predictive sample, prediction is very poor.

Thus the predictive probability of 0.447 looks a vastly optimistic and unsound statement: knowing only the mean ignores the great variability from the small sample. This should be a matter of *serious concern* to those using posterior predictive distributions for predictive probability statements.

7.3.1 Marginalization and model diagnostics

This issue can be expressed more generally, in terms of the meaning of *marginal distributions* – not just in Bayesian theory, but in general statistical theory. When we have a conditional distribution $f(y \mid x)$ of Y given $X = x$, and a

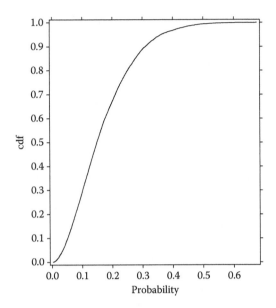

FIGURE 7.1
Beta(2,10) posterior.

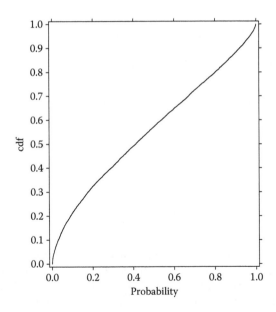

FIGURE 7.2
Binomial posterior.

marginal distribution $g(X)$ of X, the (unconditional) marginal distribution of Y is

$$h(y) = \int f(y \mid x)g(x)dx.$$

This can be viewed alternatively as the *mean* of the distribution of Y given $X = x$, averaged over X:

$$h(y) = E_X[f(y \mid X)].$$

We generally think of this marginal distribution as containing all the information about Y, after integrating out x, but from this point of view there is more information about this distribution, for example, in the *variance* of the conditional distribution averaged over X:

$$\text{Var}_X[f(Y \mid X)] = \int [f(y \mid x) - h(y)]^2 g(x)dx$$

$$= \int f^2(y \mid x)g(x)dx - h^2(y).$$

We illustrate with the posterior predictive distribution in the one-sample normal model. We have observations $y_1, \ldots, y_n$ from $N(\mu, \sigma^2)$. With flat priors on μ and $\log \sigma$, the joint posterior distribution of μ and σ can be expressed as in Chapter 2:

$$\mu \mid \sigma, \mathbf{y} \sim N(\bar{y}, \sigma^2/n), \ RSS/\sigma^2 \mid \mathbf{y} \sim \chi^2_{n-1}.$$

The predictive distribution of a new observation y_0 is

$$f(y_0 \mid \mathbf{y}) = \int \int f(y_0 \mid \mu, \sigma)\pi(\mu, \sigma \mid \mathbf{y})d\mu d\sigma$$

$$= \int \pi(\sigma \mid \mathbf{y})d\sigma \int f(y_0 \mid \mu, \sigma)\pi(\mu \mid \sigma, \mathbf{y})d\mu$$

which gives

$$y_0 \sim \bar{y} + s\sqrt{\frac{n+1}{n}} t_{n-1},$$

where $s^2 = RSS/(n-1)$. So if the normal model for Y is correct, the predictive distribution of new observations from the model should follow the scaled t_{n-1} distribution. This is frequently examined by simulating new data sets from the predictive distribution and then comparing features of the observed data with those of the simulated data sets. There are many ways of doing this; one would be to construct the envelope of the empirical cdfs from the simulated data sets and see whether this envelope contains the empirical cdf of the observed data.

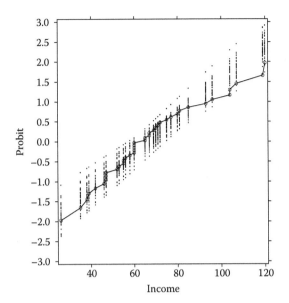

FIGURE 7.3
Posterior draws of normal cdf, income sample.

This envelope approach formalizes the common approach of plotting the empirical cdf of the data together with the cdf of the predictive *t*-distribution: if this comparison is appropriate the agreement should be good. Of course, we would like to plot the cdf of the *true* (or nearest-to-true) normal distribution, but substituting the MLEs in the normal cdf will not be sufficient: this maximizes the agreement between model and data.

However, it should be clear from the binomial discussion above that something is being missed: the variability in the posterior distribution of the model parameters is not being correctly represented in the predictive *t*-distribution. We can allow for this by making M draws from the joint posterior of μ and σ and substituting them in the normal cdf $\Phi[(y - \mu)/\sigma]$, to give M draws $\Phi[(y - \mu^{[m]})/\sigma^{[m]}]$. The envelope of these draws should contain the empirical cdf of the observed data if the normal model is adequate.

We illustrate these two approaches with a sample from a skewed distribution (the income population of Chapter 2). Figure 7.3 shows, on the probit scale, the envelope of 39 draws for the StatLab income sample.

If the normal model is correct, the probability that the empirical cdf at any income point falls outside the envelope of the 39 draws is 0.05 (0.025 at each end), so the envelope is a 95% pointwise confidence band for the normal cdf (Atkinson 1985). The empirical cdf falls outside the edge of the envelope at income 119, and is right on the edge at 104. This suggests failure of the normal model.

Figure 7.4 shows, also on the probit scale, the envelope of the empirical cdfs of 39 samples of size 40 drawn from the random normal distributions for new

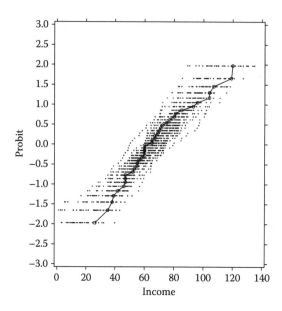

FIGURE 7.4
Posterior draws from predictive distribution, income sample.

data: the mth sample is generated from the normal distribution $N(\mu^{[m]}, \sigma^{[m]2})$. The graph scales and bands are different because the variability is across the data values which vary from sample to sample, rather than across the cdf for the given sample.

There is a striking difference in the envelope variability in this graph compared with the posterior cdf distribution in the first graph: the normal distributions for new data give a much greater range for the empirical cdfs. The empirical cdf of the income sample falls comfortably near the centre of the envelope, and there is no evidence of any departure from the assumed normal model in the graph.

A formal Bayesian treatment of the envelope interpretation is not given here: the point is to illustrate the much greater variability in the posterior predictive distribution approach which raises similar concerns to those expressed by Hjort et al. (2006).

The approach using the posterior distribution of the cdf can be extended quite generally. For a model density $f(y \mid \theta)$ with multidimensional θ (for example, a regression model), we observe response values y_i and explanatory variable values x_i. Given a prior distribution $\pi(\theta)$, we make M random draws $\theta^{[m]}$ from the posterior for θ, and substitute them in the density to give M draws from the density $f(y_i \mid \theta^{[m]})$ for each i. The set of M empirical cdfs $F(y_i \mid \theta^{[m]})$ define an envelope, and by choosing the percentiles of the simulation distribution appropriately, we can construct a pointwise $100(1 - \alpha)\%$ confidence region for the true cdf. If this does not contain the observed data empirical cdf, we have evidence against the assumed distributional model.

We do not give further details here.

7.4 Multinomial deviance computation

We now consider the comparison of multinomial and parametric models through their posterior deviance distributions. For the multinomial, we use the Haldane prior, giving the Dirichlet posterior on the observed support as described in Chapter 4.

For the normal or other parametric model, given as $f(y \mid \theta)$, we need care in computing the corresponding deviance. Let δ be the measurement precision, so that a recorded value of y corresponds to a true value of $Y \in (y + \delta/2, y - \delta/2)$. The likelihood is formally

$$L(\theta \mid \mathbf{y}) = \prod_{i=1}^{n} [F(y_i + \delta/2 \mid \theta) - F(y_i - \delta/2 \mid \theta)].$$

If the measurement precision of Y is high, δ is small compared with the variability of the distribution (σ for the normal), and the likelihood can then be approximated accurately by

$$L^*(\theta \mid \mathbf{y}) = \prod_{i}^{n} [[f(y_i \mid \theta) \cdot \delta].$$

In this case, the usual likelihood results and the standard diffuse priors give the standard posterior distribution for θ. However, if δ is *large* (low-measurement precision), then the likelihood and posteriors have to be computed from the "grouped data" likelihood given exactly above as the product of differences between cdfs. In either case, we generate M values $\theta^{[m]}$ from the posterior distribution of θ, and use these in the appropriate (grouped or ungrouped) model likelihood.

How we use them requires further care. For continuous distributions, *every* value of Y, observed or unobserved, has positive probability under the model. However, under the multinomial distribution with the Haldane prior, only *observed* values have positive probability. A simple comparison of

$$L(\theta \mid \mathbf{y}) = \delta^n \cdot \prod_{i=1}^{n} f(y_i \mid \theta)$$

with the above multinomial likelihood will unduly favor the multinomial model because of the assignment by the continuous model of positive probability to the unobserved values. To correct this, the unobserved values *also* have to be assigned zero probability under the continuous model. This is simply achieved by scaling up to 1.0 the sum of the continuous model probabilities on the observed data: we write

$$q_j^{[m]} = F\left(y_j + \delta/2 \mid \theta^{[m]}\right) - F\left(y_j - \delta/2 \mid \theta^{[m]}\right)$$

$$q_+^{[m]} = \sum_{j=1}^{d} q_j^{[m]}$$

$$p_j^{[m]} = q_j^{[m]}/q_+^{[m]}$$

$$D_{cts}^{[m]} = -2\sum_{j=1}^{d} n_j \log p_j^{[m]}$$

$$= -2\sum_{j=1}^{d} n_j \log \left[q_j^{[m]}/q_+^{[m]}\right]$$

$$= -2\left[\sum_{j=1}^{d} n_j \log q_j^{[m]} - n \log q_+^{[m]}\right],$$

a *corrected* or *penalized* form of the continuous model deviance which accounts for the constraint scaling of the model probabilities. Now the distribution of the continuous model deviance D_{cts} is comparable with that of the multinomial model D_{mult}.

An obvious question here is how the measurement precision affects the model comparison. If we increased the measurement precision, for example, by recording an extra decimal place, what effect would this have?

This is easily assessed. If the measurement precision is high, the continuous model deviance expression above has the simpler form:

$$q_j^{[m]} = F\left(y_j + \delta/2 \mid \theta^{[m]}\right) - F\left(y_j - \delta/2 \mid \theta^{[m]}\right)$$

$$= f\left(y_j \mid \theta^{[m]}\right)\delta$$

$$q_+^{[m]} = \sum_{j=1}^{d} f\left(y_j \mid \theta^{[m]}\right)\delta$$

$$p_j^{[m]} = f\left(y_j \mid \theta^{[m]}\right)/\sum_{j=1}^{d} f\left(y_j \mid \theta^{[m]}\right)$$

$$D_{cts}^{[m]} = -2\sum_{j=1}^{d} n_j \log p_j^{[m]}$$

$$= -2\left[\sum_{j=1}^{d} n_j \log f\left(y_j \mid \theta^{[m]}\right) - n \log\left(\sum_{j=1}^{d} f\left(y_j \mid \theta^{[m]}\right)\right)\right],$$

which does not depend at all on δ. So increasing the measurement precision, no matter how finely, does not change the normal model deviance. *Reducing the measurement precision, by coarsening the data recording, may* affect the deviance, but this is to be expected since it represents a loss of information.

7.5   Model comparison through posterior deviances

The two sets of deviances can be used directly to determine the relative support for the two models from the data. For each m, the deviances $D_{cts}^{[m]}$ and $D_{mult}^{[m]}$ are deviances for *fully specified* continuous and multinomial models, respectively (the continuous model being truncated to the observed support). So the corresponding model likelihoods can be converted to a likelihood ratio

$$LR^{[m]} = \frac{\exp\left(-D_{cts}^{[m]}/2\right)}{\exp\left(-D_{mult}^{[m]}/2\right)}.$$

By the usual calibration of likelihood ratios for fully specified models, the support for the null continuous model over the alternative multinomial model is moderate if $LR > 9$, while an $LR > 1$ gives a simple preference for the continuous model. The likelihood ratio can be converted to posterior odds on the models, given the prior odds. An equivalent computation uses the deviance differences: if

$$D_{cts}^{[m]} - D_{mult}^{[m]} < -2\log 9,$$

we have moderate sample support for the continuous model.

Over the M draws we have a distribution of the support strength for the continuous model; if a high proportion – say 90% – of the deviance differences are negative, we have posterior probability approximately 0.9 that the deviance of the continuous model is smaller than that of the multinomial – that is, that the continuous model is better supported than the multinomial.

If, however, this proportion is around 50%, we do not have sufficient evidence to choose between the two models. In this case the parametric model can be considered no worse than the multinomial model, so is an acceptable candidate for a parsimonious representation. (This does not exclude the possibility that other parametric models may fit equally well, or better.)

If the proportion of negative deviances is 10% or less, we have strong evidence that the multinomial model is better supported than the continuous model – that is, that the continuous distribution is mis-specified.

We use throughout this chapter equal prior probabilities π_j on all competing models, but there is no difficulty in incorporating nonuniform model priors, by a location shift of $-2\log \pi_j$ in the deviance distribution for model j.

It may seem surprising that no penalty is needed for the heavily parametrized multinomial compared with the parsimonious continuous distribution. The penalty is in fact already implicit in the distribution of the deviance, for the information in the data is spread very thinly for the Dirichlet, since the sample size at each data point determines the precision of the multinomial parameter at that data point. Thus the distribution of the multinomial deviance is much more diffuse than that of the continuous model deviance. This point is illustrated by the asymptotic distribution of the model deviance:

$$-2\log L(\theta) \rightarrow -2\log L(\hat\theta) + \chi_p^2$$

where p is the number of model parameters. With increasing parameter dimension the χ_p^2 distribution has increasing variance: on the cdf scale its "slope" increases with p, as we show below in the examples.

7.6 Examples

We reanalyze several examples from Conigliani et al. (2000).

7.6.1 Three binomial models

Conigliani et al. (2000) gave three data sets to test whether the observations come from a binomial distribution (their Example 1). They compared the Pearson chi-square test with the fractional Bayes factor (for the alternative multinomial model against the null binomial model, Table 7.1).

The approaches appear to be in agreement for the two extremes, less so for the intermediate case where the p-value from the X^2 test would be 0.0065 if the χ_1^2 distribution were appropriate, strong evidence against the null binomial, whereas the FBF, if converted to a posterior probability of the null hypothesis, would be 0.1534.

For the posterior distribution of the likelihood, we use the Haldane prior under each model. The posterior distributions of both deviances are shown as cdfs in Figures 7.5, 7.6, and 7.7, corresponding to the three data sets (solid curve, binomial; dashed curve, multinomial).

The posterior distributions of the deviance *differences* (multinomial-binomial) are shown in Figures 7.8, 7.9, and 7.10.

TABLE 7.1

Data, X^2, and fractional Bayes factor

y	n_1	X^2	FBF	n_2	X^2	FBF	n_3	X^2	FBF
0	64	29.63	71839	16	7.41	5.52	4	1.85	0.96
1	16			4			1		
2	16			4			1		
	96			24			6		

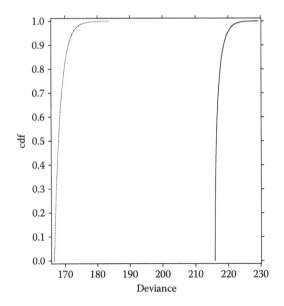

FIGURE 7.5
Deviances: $n = 96$.

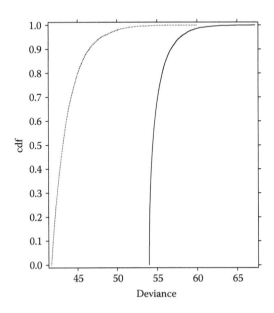

FIGURE 7.6
Deviances: $n = 24$.

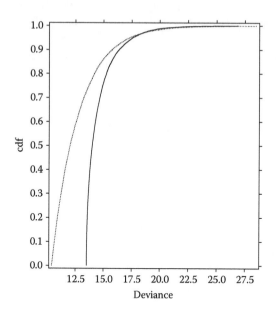

FIGURE 7.7
Deviances: $n = 6$.

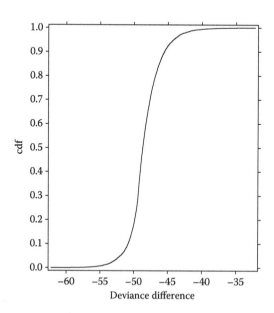

FIGURE 7.8
Deviance difference: $n = 96$.

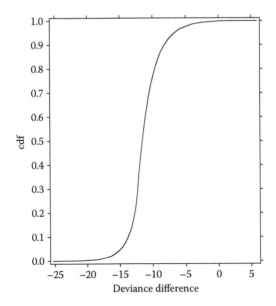

FIGURE 7.9
Deviance difference: $n = 24$.

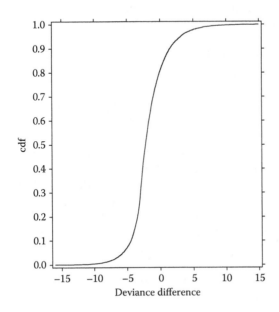

FIGURE 7.10
Deviance difference: $n = 6$.

The 95% credible intervals for the true deviance differences for each data set are $[-42.96, -53.12]$, $[-5.08, -16.06]$, and $[4.67, -6.89]$, respectively. Converting these to likelihood ratios for multinomial to binomial, comparable to the FBF, we have $[3.43 \times 10^{11}, 2.13 \times 10^9]$, $[12.68, 3072]$, and $[0.097, 31.34]$. The first two intervals do not include the corresponding FBF, which is much smaller: the FBF for these examples gives weaker evidence than the likelihood ratio distribution. For the intermediate sample size the posterior likelihood ratio conclusion agrees with the p-value conclusion: there is quite strong evidence against the binomial model.

7.6.2   Poisson model

```
---------------------------------------------
y    0       1      2      3     4     5     6
n    156     63     29     8     4     1     1
m    135.89  89.21  29.28  6.41  1.05  0.14  0.02
---------------------------------------------
Number y of times "may" appears per block,
number n of blocks, and fitted Poisson
frequencies m
```

A second example from Conigliani et al. (2000) is of data possibly from a Poisson model (their Example 5). The data from Hoaglin et al. (1985) are the number of times the word "may" appears per block in papers by James Madison. They are given above with the fitted frequencies from the Poisson model.

The fit is clearly very poor, and the Pearson X^2 is 15.03 on grouping the last four cells, strong evidence against the Poisson model ($p = 0.0018$ from χ_3^2). The fractional Bayes factor is only 1.62, very weak evidence against the null model. *All* the other frequentist test criteria given by Conigliani et al. (2000) give the same result: that the Poisson model is strongly rejected.

For the posterior likelihood ratio, we use the Haldane prior for the multinomial model and the flat prior on $\log \mu$ for the Poisson mean μ. (The choice of Poisson prior has no effect on the conclusions – any diffuse prior gives the same conclusions.) The posterior distributions of both deviances are shown as cdfs in Figure 7.11, and the posterior distribution of the deviance *difference* (multinomial – Poisson) is shown in Figure 7.12. *All* of the 10,000 simulated deviance differences are negative, and the 95% credible interval for the true difference is $[-442.4, -458.2]$. The evidence against the Poisson is overwhelming.

We now describe a simulation study to check how well this procedure works with continuous data.

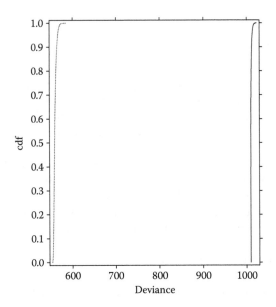

FIGURE 7.11
Deviances: Madison data.

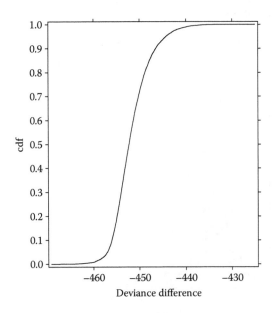

FIGURE 7.12
Deviance difference: Madison data.

7.7   Simulation study

We assess the method in a simulation of data with increasing sample sizes from one of four populations. These populations are based on the StatLab database (Hodges et al. 1975). We use the subpopulation of 648 families with a boy, and the annual family income at birth (recorded in 100s of 1961 dollars) is the variable of interest. The histogram of counts at each distinct value is shown in Figure 7.13.

The distribution is heavily skewed, a common feature of income data. The cdf of the population income values is shown on the normal, lognormal, and gamma deviate scales in Figures 7.14, 7.15, and 7.16 with the corresponding distributions fitted by maximum likelihood. The gamma distribution fits well, except at both extremes, while the lognormal fits less well, with the wrong curvature.

The four populations used for simulation are

- The actual family incomes from the StatLab boy subpopulation
- A normal distribution with the same mean and variance as the StatLab family income subpopulation
- A lognormal distribution with the same mean and variance as the StatLab family income subpopulation
- A gamma distribution with the same mean and variance as the StatLab family income subpopulation

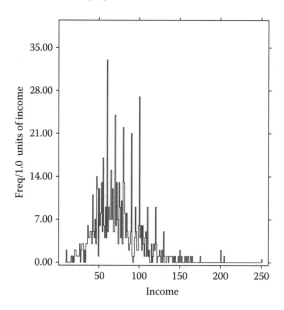

FIGURE 7.13
Income histogram.

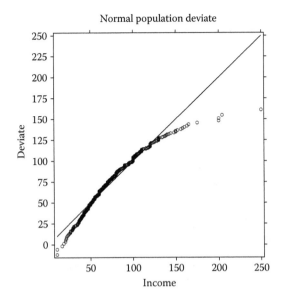

FIGURE 7.14
Income population: normal deviate scale.

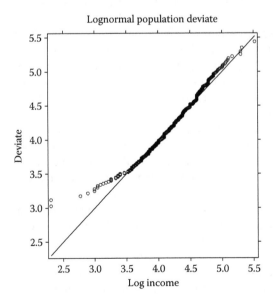

FIGURE 7.15
Income population: lognormal deviate scale.

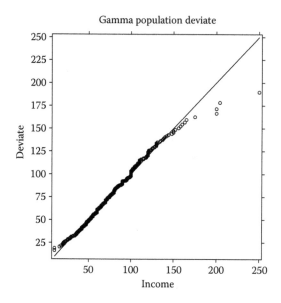

FIGURE 7.16
Income population: gamma deviate scale.

For the three continuous distributions, the generated values of family income are truncated to integer values to correspond to those from the real StatLab population; the measurement precision δ is thus 1 unit.

We report four studies, in each of which the data are generated by sampling with replacement from one of the above population models. For each data set, we generate $M = 10,000$ draws from the posterior distribution of the deviance for each of the four distributions: multinomial, normal, lognormal, and gamma. For the continuous distributions, the model probabilities are scaled to sum to 1.0 over the observed data.

For each draw, we find the distribution with the largest likelihood (smallest deviance) and compute over the 10,000 draws the "best model" proportions of draws in which each distribution has the smallest deviance. (These are not the same as the posterior probabilities of choosing the models, but correspond to a forced choice of the "best" model at each draw.)

We repeat the sample generation and posterior probability computation over 1000 independent samples, and repeat the whole simulation for a set of increasing sample sizes n. The averages over the 1000 samples of the best model proportions are shown in Tables 7.2–7.5.

Several characteristic features of this model selection process are visible in all the tables. For very small samples (10–40) the normal distribution is chosen most often, regardless of the true distribution. In small samples (40–100) from the parametric distributions the multinomial has consistently low probability of being chosen – the multinomial support probabilities are so poorly defined that its deviance distribution is shifted to the right relative

TABLE 7.2

Best Model Proportions: True Normal Distribution

n	Normal	Lognormal	Gamma	Multinomial
10	0.4015	0.3304	0.2139	0.0542
20	0.5200	0.2426	0.2073	0.0301
30	0.6319	0.1503	0.1973	0.0205
40	0.6784	0.1066	0.1973	0.0176
50	0.6991	0.0845	0.2003	0.0160
60	0.7336	0.0635	0.1843	0.0186
70	0.7425	0.0594	0.1780	0.0201
80	0.7805	0.0352	0.1610	0.0233
90	0.7917	0.0473	0.1345	0.0265
100	0.7949	0.0407	0.1330	0.0314
200	0.8095	0.0218	0.0308	0.1379
300	0.7307	0.0035	0.0069	0.2589
400	0.6736	0.0011	0.0029	0.3224
500	0.6440	0.	0.0016	0.3543
600	0.6352	0.	0.0004	0.3644
700	0.6011	0.	0.	0.3989
800	0.6064	0.	0.	0.3936
900	0.5710	0.	0.	0.4290
1000	0.5672	0.	0.	0.4328

TABLE 7.3

Best Model Proportions: True Lognormal Distribution

n	Normal	Lognormal	Gamma	Multinomial
10	0.4227	0.2417	0.2778	0.0579
20	0.3601	0.2942	0.3121	0.0337
30	0.2909	0.3270	0.3552	0.0270
40	0.2343	0.3459	0.3961	0.0237
50	0.1799	0.3704	0.4249	0.0247
60	0.1313	0.3932	0.4479	0.0276
70	0.1016	0.4017	0.4640	0.0326
80	0.0717	0.4234	0.4643	0.0406
90	0.0511	0.4324	0.4692	0.0474
100	0.0413	0.4362	0.4657	0.0567
200	0.0031	0.4580	0.3305	0.2084
300	0.0028	0.4810	0.1968	0.3194
400	0.0008	0.4987	0.1064	0.3940
500	0.0001	0.5343	0.0609	0.4047
600	0.	0.5436	0.0361	0.4203
700	0.	0.5573	0.0187	0.4241
800	0.	0.5348	0.0111	0.4541
900	0.	0.5474	0.0044	0.4482
1000	0.	0.5324	0.0043	0.4632

TABLE 7.4

Best Model Proportions: True Gamma Distribution

n	Normal	Lognormal	Gamma	Multinomial
10	0.4203	0.2638	0.2605	0.0553
20	0.3976	0.2861	0.2864	0.0299
30	0.3752	0.2711	0.3336	0.0201
40	0.3593	0.2588	0.3649	0.0170
50	0.2957	0.2760	0.4115	0.0168
60	0.2630	0.2811	0.4374	0.0185
70	0.2400	0.2723	0.4670	0.0207
80	0.2272	0.2642	0.4836	0.0250
90	0.1976	0.2750	0.4975	0.0298
100	0.1871	0.2647	0.5095	0.0386
200	0.0493	0.2408	0.5640	0.1459
300	0.0122	0.1987	0.5328	0.2563
400	0.0029	0.1731	0.5105	0.3135
500	0.0001	0.1562	0.5073	0.3363
600	0.	0.1384	0.5086	0.3530
700	0.	0.1084	0.5098	0.3818
800	0.	0.0977	0.5128	0.3895
900	0.	0.1029	0.5045	0.3926
1000	0.	0.0887	0.5073	0.4040

TABLE 7.5

Best Model Proportions: True Multinomial Distribution

n	Normal	Lognormal	Gamma	Multinomial
10	0.4406	0.2457	0.2468	0.0669
20	0.4441	0.2264	0.2820	0.0474
30	0.4010	0.2148	0.3326	0.0516
40	0.3862	0.2005	0.3552	0.0582
50	0.3746	0.1783	0.3712	0.0759
60	0.3185	0.1849	0.3865	0.1100
70	0.2994	0.1685	0.3874	0.1447
80	0.2912	0.1440	0.3723	0.1924
90	0.2616	0.1286	0.3493	0.2605
100	0.2056	0.1166	0.3413	0.3366
120	0.1546	0.0783	0.2536	0.5135
140	0.0853	0.0490	0.1711	0.6946
160	0.0415	0.0258	0.1015	0.8312
200	0.0072	0.0057	0.0258	0.9614
300	0.	0.	0.0001	0.9999

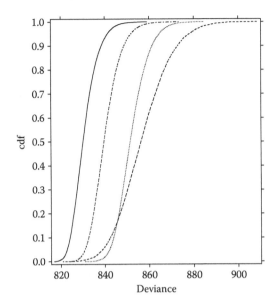

FIGURE 7.17
Deviances: true normal, $n = 100$.

to the others. Figure 7.17 shows this for a sample of 100 from the normal distribution. (The curves are, at top from left, normal, gamma, lognormal, multinomial.)

As the sample size becomes large, the choice probabilities of the incorrect parametric models go to zero, while those for the true parametric model *and* the multinomial model approach 0.5. This reflects the form of the deviance distribution for the multinomial: as the sample size increases, so does the sample support, but more slowly. So this deviance distribution becomes more diffuse (from the increased number of parameters) but better defined (as the sample sizes on the support points increase), and since it is by definition a "correct" model, its deviance distribution crosses that for the true parametric model around the median of each distribution. Figure 7.18 shows this property for a sample of 1000 from the gamma distribution. (The curves are, at bottom from left, multinomial, gamma, lognormal, normal.)

This does not create any difficulty for model choice, as it is clear that the gamma distribution (in this example) is no worse than the multinomial, and is therefore an acceptable parametric model. For the true parametric distributions, the normal distribution is fairly easily identified: at sample size 100 it has average probability 0.8 of being the "best." The true lognormal is much harder to identify, the gamma providing a more plausible alternative up to sample size 100. The true gamma distribution is easier to distinguish from the lognormal – it has the highest choice probability by sample size 40.

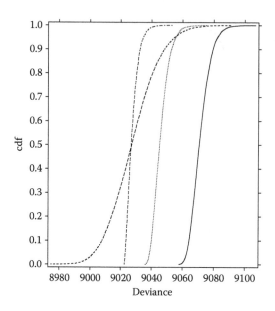

FIGURE 7.18
Deviances: true gamma, $n = 1000$.

7.8 Discussion

Bayes factors have been the traditional Bayesian tool for model comparisons. In the case of goodness of fit testing for an assumed probability model, Bayes factors cannot be used without proper informative priors because an improper prior leads to an undefined integrated likelihood. With a proper informative prior, the integrated likelihood depends *strongly* on the prior parameters, and this dependence does not dissipate with increasing sample size (Kass and Raftery 1995).

Ad hoc attempts to deal with this problem through variations of the Bayes factor, whether by fractionating the sample into training and validation subsamples (the fractional Bayes factor of O'Hagan 1995 and the intrinsic Bayes factor of Berger and Pericchi 1996) or by integrating with respect to the posterior (the posterior Bayes factor of Aitkin 1991), cannot be calibrated in the same way as Bayes factors. The fractional Bayes factor does not perform well for continuous data.

By mapping the parameter uncertainty into the likelihood, we are able to retain the usual calibration of likelihood ratios for fully specified models and so are able to compare models without informative prior specifications or arbitrary assignment of training sample fractions.

The behavior of the model choice criterion resulting from this approach is different from that expected from Bayes factors or the Bayesian information criterion (BIC). An appealing feature of Bayesian model choice has always been that it can identify a true model with probability 1 as the sample size goes to infinity. For the comparisons of parsimonious parametric models, this occurs also for the posterior likelihood ratio criterion, but in the comparisons of the multinomial with parsimonious parametric models, the behavior of the criterion is different: as n increases the probability for the true model approaches 0.5, as does that for the multinomial model. This is not unreasonable, since the multinomial model is also correct, though heavily parametrized. The BIC and Bayes factors have explicit or implicit penalties on the model parametrization which would counteract the deviance, but this is not needed for us to come to the correct conclusion: that the parametric model represents the data just as well as the multinomial model, but much more precisely, and can therefore serve as a parsimonious model for the data. In this respect the likelihood ratio comparison performs like the frequentist test of a parsimonious model against the "saturated" model.

An implicit "penalty" arises from the diffuseness of the deviance distribution for the multinomial, and this increases with sample size, so that it is only for the *true* multinomial model that the multinomial deviance becomes clearly superior to the parametric model deviances when the sample size increases.

8

Complex Models

In this chapter we consider two classes of complex models: two-level variance component models and finite mixture models. These models require in general iterative Markov chain Monte Carlo (MCMC) methods to obtain the necessary posterior distributions for the model parameters and likelihood or deviance. There are extensive discussion of these methods in many texts, for example, Robert and Casella (1999).

8.1 The data augmentation algorithm

We give only the briefest sketch of the data augmentation (DA) algorithm (Tanner and Wong 1987, Tanner 1996), which can be used for an *incomplete data* formulation of the model, when this is available. This is a stochastic version of the expectation-maximization (EM) algorithm; if the model $f(y \mid \theta, \phi) = f(y \mid \lambda)$ can be expressed in terms of a *latent variable* z, in the form

$$f(y \mid \lambda) = \int f(y \mid z, \theta)g(z \mid \phi)dz,$$

then it has an EM analysis in terms of iterating between the maximization of the *complete data log-likelihood* of y *and* z (the M step) and the replacement of terms involving z in the complete data likelihood by their *conditional expectations* (the E step) given λ and y.

The DA analysis replaces the M step by a *posterior simulation* step of drawing from the complete data posterior of λ given z and y, $\pi(\lambda \mid z, y)$, and replaces the E step by a *data augmentation* step, of drawing from the posterior of z given λ and y, $\pi(z \mid \lambda, y)$.

The convergence of the EM algorithm to the MLE of λ, and of the conditional expectation of z, is replaced by the convergence *in distribution* of λ to its marginal posterior distribution given y, and of z to its marginal posterior distribution given y.

A noniterative solution for the posterior distribution of λ can be obtained in some models from the *inverse Bayes formula* (IBF) given by Ng (1997) and discussed in detail by Tan et al. (2010). The basis of the solution is a rewriting of Bayes's theorem relating the conditional distributions of λ given z and y, and z given λ and y:

$$f(\lambda \mid z, y) f(z \mid y) = f(z \mid \lambda, y) f(\lambda \mid y)$$

$$f(z \mid y) = \frac{f(z \mid \lambda, y)}{f(\lambda \mid z, y)} f(\lambda \mid y)$$

$$1 = \int f(z \mid y) dz = \int \left[\frac{f(z \mid \lambda, y)}{f(\lambda \mid z, y)} \right] dz f(\lambda \mid y)$$

$$f(\lambda \mid y) = \left\{ \int \left[\frac{f(z \mid \lambda, y)}{f(\lambda \mid z, y)} \right] dz \right\}^{-1}$$

$$= \left[\int f^{-1}(\lambda \mid z, y) f(z \mid \lambda, y) dz \right]^{-1}$$

If both the conditional distributions of $\lambda \mid z, y$ and $z \mid \lambda, y$ are analytic, it may be possible to integrate the reciprocal of the conditional posterior of λ over z, with respect to the posterior distribution of z given λ, to obtain an analytic expression for the posterior of λ which avoids MCMC iteration altogether. Tan et al. (2010) give many examples of analytic posteriors which can be obtained in this way. We comment on this for the finite mixture example below. We do not discuss MCMC computational methods further, nor give details of Winbugs or other MCMC packages.

8.2   Two-level variance component models

In Chapter 4, we considered a simple balanced two-level data set from the book by Box and Tiao (1973), who gave a full exposition of Bayesian analysis assuming normal distributions for the response (grams of standard color) and the batch random effects. Their analysis used analytic approximations; we use instead the simulation approach to obtain the posteriors for the various quantities of interest.

Box and Tiao (1973) gave a small example of a designed experiment in which five samples were randomly chosen from six batches of raw material, and a laboratory determination made (of the yield of dyestuff in grams of standard color) on each sample. We want to use the *batch variation* in color to *improve the precision* of the individual batch means: batches are the first-stage sampling level, and samples within batch are the second stage.

We reproduce the data table from Chapter 4 (Table 8.1).

TABLE 8.1

Dyestuff Data

Batch	1	2	3	4	5	6
	1545	1540	1595	1445	1595	1520
	1440	1555	1550	1440	1630	1455
	1440	1490	1605	1595	1515	1450
	1520	1560	1510	1465	1635	1480
	1580	1495	1560	1545	1625	1445
Mean	1505.0	1528.0	1564.0	1498.0	1600.0	1470.0
variance	3975.0	1107.5	1442.5	4720.0	2500.0	962.5

8.2.1 Two-level fixed effects model

We model the yield in grams as normal: the *fixed effects* model is

$$y_{ij} \mid \mu_i \sim N(\mu_i, \sigma^2), \; j = 1, \ldots, n_i, \; i = 1, \ldots, r$$

where the μ_i are *unstructured* mean parameters.

Bayes inference for the mean parameters for this model is the same as that for a single sample, except that the variance inference is based on the *pooled within-sample variability* $W = \sum_i \sum_j (y_{ij} - \bar{y}_i)^2$, with degrees of freedom $v = \sum_i (n_i - 1)$. So

$$W/\sigma^2 \sim \chi_v^2,$$
$$s^2 = W/v,$$
$$\mu_i \mid \text{data} \sim \bar{y}_i + T_v s / \sqrt{n_i}.$$

where T_v has the t distribution with v degrees of freedom.

8.2.2 Two-level random effects model

The *random effect* model adds the model of normal variation in the means across batches: $\mu_i \sim N(\mu, \sigma_B^2)$, with variance components σ^2 and σ_B^2. It follows immediately that

$$\bar{y}_i \mid \mu_i \sim N(\mu_i, \sigma^2/n_i),$$
$$\bar{y}_i \sim N(\mu, \phi_i^2),$$
$$\mu_i \mid \bar{y}_i \sim N(w_i \bar{y}_i + (1 - w_i)\mu, \sigma_B^2(1 - w_i))$$

where

$$\phi_i^2 = \sigma_B^2 + \sigma^2/n_i, \; \theta = \sigma_B^2/\sigma^2,$$
$$w_i = n_i \theta / (1 + n_i \theta) = 1 - \sigma^2/(n_i \phi_i^2).$$

8.2.3   Posterior inference

The posterior mean of μ_i is used in *empirical Bayes* estimation as a *shrinkage estimator* of the batch mean μ_i. This approach is widely used, with MLEs replacing the unknown variance component parameters and overall mean:

$$\text{E}[\widehat{\mu_i \mid \bar{y}_i}] = \widehat{w}_i \bar{y}_i + (1 - \widehat{w}_i)\hat{\mu}$$
$$\text{Var}[\widehat{\mu_i \mid \bar{y}_i}] = \hat{\sigma}_B^2(1 - \widehat{w}_i)$$
$$\widehat{w}_i = 1 - \hat{\sigma}^2/(n_i\hat{\phi}_i^2)$$
$$\mu_i \mid \bar{y}_i \approx N(\text{E}[\widehat{\mu_i \mid \bar{y}_i}], \text{Var}[\widehat{\mu_i \mid \bar{y}_i}]).$$

The true variability of the resulting shrinkage estimator is very difficult to establish, and the *variances* of the $\mu_i \mid \bar{y}_i$ are seriously underestimated.

We need the *full posterior distribution* of the $\mu_i \mid \bar{y}_i$, allowing for *all* the sources of uncertainty.

8.2.4   Likelihood

We deal with only the simplest case: $n_i = n$ for all i. The likelihood is found by integrating out the unobservable μ_i:

$$L(\mu, \sigma, \sigma_B) = \prod_{i=1}^{r} \int \left[\prod_{j=1}^{n} f(y_{ij} \mid \mu_i) \right] g(\mu_i) d\mu_i$$

$$\prod_{j=1}^{n} f(y_{ij} \mid \mu_i) = c \cdot \prod_{j=1}^{n} \frac{1}{\sigma} \exp \left\{ -\frac{1}{2\sigma^2} (y_{ij} - \mu_i)^2 \right\}$$

$$= c \cdot \frac{1}{\sigma^n} \exp \left\{ -\frac{1}{2\sigma^2} \sum_{j=1}^{n} (y_{ij} - \bar{y}_i + \bar{y}_i - \mu_i)^2 \right\}$$

$$= c \cdot \frac{1}{\sigma^{n-1}} \exp \left\{ -\frac{W_i}{2\sigma^2} \right\} \cdot \frac{1}{\sigma} \exp \left\{ -\frac{1}{2\sigma^2} n(\bar{y}_i - \mu_i)^2 \right\}$$

where $W_i = \sum_{j=1}^{n} (y_{ij} - \bar{y}_i)^2$. Then

$$L(\mu, \sigma, \sigma_B) = c \cdot \frac{1}{\sigma^{r(n-1)}} \exp \left\{ -\frac{W}{2\sigma^2} \right\}$$

$$\cdot \prod_{i=1}^{r} \int \frac{1}{\sigma} \exp \left\{ -\frac{1}{2\sigma^2} n(\bar{y}_i - \mu_i)^2 \right\} \cdot \frac{1}{\sigma_B} \exp \left\{ -\frac{1}{2\sigma_B^2} (\mu_i - \mu)^2 \right\} d\mu_i$$

$$= c \cdot \frac{1}{\sigma^{r(n-1)}} \exp \left\{ -\frac{W}{2\sigma^2} \right\} \cdot \prod_{i=1}^{r} \frac{1}{\phi} \exp \left\{ -\frac{1}{2\phi^2} (\bar{y}_i - \mu)^2) \right\}$$

$$= c \cdot \frac{1}{\sigma^{r(n-1)}} \exp \left\{ -\frac{W}{2\sigma^2} \right\} \cdot \left(\frac{1}{\phi} \right)^{r-1} \cdot \exp \left\{ -\frac{B}{2\phi^2} \right\}$$

$$\cdot \frac{1}{\phi} \exp \left\{ -\frac{r}{2\phi^2} (\bar{y} - \mu)^2 \right\}$$

where

$$\bar{y} = \sum_i \bar{y}_i / r, \quad W = \sum_i W_i, \quad B = n \sum_i (\bar{y}_i - \bar{y})^2, \quad \phi^2 = \sigma_B^2 + \sigma^2/n.$$

8.2.5 Maximum likelihood estimates

We have the frequentist results

- $\bar{y} \sim N(\mu, \phi^2/r)$, $\hat{\mu} = \bar{y}$
- $B/\phi^2 \sim \chi^2_{r-1}$, $\hat{\phi}^2 = B/(r-1)$
- $W/\sigma^2 \sim \chi^2_{r(n-1)}$, $\hat{\sigma}^2 = W/[r(n-1)]$

The MLE of σ_B^2 is then $\hat{\sigma}_B^2 = \hat{\phi}^2 - \hat{\sigma}^2/n$ *unless the RHS is negative*, in which case $\hat{\sigma}_B^2 = 0$. (It is occasionally argued in the frequentist framework that negative estimates should be allowed, since the above nonnegative estimator is biased for σ_B^2, whereas the unrestricted estimator which can be negative is unbiased. In the Bayesian framework bias is irrelevant, and nonnegative parameters should not have negative estimates.

A separate issue is the possibility of *negative intraclass correlation* of the within-batch responses, as may occur in plant competition experiments; this requires an extended model in which the within-batch responses are *conditionally dependent* given the random effects.)

For example, we have $r = 6$, $n = 5$, $\bar{y} = 1527.5$, $W = 58,830$, $B = 11,271.5$. So

$$\hat{\sigma}^2 = 2,451.25, \quad \hat{\phi}^2 = 2,254.3, \quad \hat{\sigma}_B^2 = 1,764.05.$$

Among-batch variation in the batch *mean* yields is nearly as large as the within-batch variation of *individual* yields.

For the empirical Bayes (EB) estimates, we have

$$\hat{\theta} = \hat{\sigma}_B^2/\hat{\sigma}^2 = 0.720$$
$$\hat{w}_i = 0.783 \text{ for all } i$$
$$\widehat{\mu}_i = 0.783\bar{y}_i + 0.217 \times 1527.5$$
$$\widehat{\text{Var}[\mu_i]} = 1,764.05 \times 0.217 = 382.8.$$

Graphs of empirical Bayes posteriors for each batch mean are shown later.

8.2.6 Posteriors

With flat priors on μ, $\log \sigma$ and $\log \phi$, we have

- $\mu \mid \bar{y}, \phi \sim N(\bar{y}, \phi^2/r)$
- $B/\phi^2 \sim \chi^2_{r-1}$, independently of
- $W/\sigma^2 \sim \chi^2_{r(n-1)}$

These are direct analogues of the frequentist results. However, the parameter spaces of σ^2 and ϕ^2 are *related*, because $\phi^2 = \sigma_B^2 + \sigma^2/n \geq \sigma^2/n$. We need to ensure that the parameter draws are *restricted* to the legal parameter space. We draw the parameters sequentially. Set $m = 1$, then

- Make a random draw u from $\chi^2_{r(n-1)}$, and calculate $\sigma^{[m]2} = W/u$.
- Make an independent random draw v from χ^2_{r-1}, and calculate $\phi^{[m]2} = B/v$.
- If $\phi^{[m]2} \leq \sigma^{[m]2}/n$, discard u and v, make new random draws u and v and repeat the above steps.
- If $\phi^{[m]2} > \sigma^{[m]2}/n$,
  - Set $\sigma_B^{[m]2} = \phi^{[m]2} - \sigma^{[m]2}/n$.
  - Evaluate $w_i^{[m]} = 1 - \sigma^{[m]2}/(n\phi^{[m]2})$.
  - Make one random draw $\mu^{[m]}$ from $N(\bar{y}, \phi^{[m]2}/r)$.
  - Make one random draw $\mu_i^{[m]}$ from
    $N(w_i^{[m]}\bar{y}_i + [1 - w_i^{[m]}]\mu^{[m]}, \sigma_B^{[m]2}[1 - w_i^{[m]}])$.
- Set $m = m + 1$.

8.2.7   Box-Tiao example

We show the full posterior distribution of the variance components in Figure 8.1 (solid, σ^2; dotted, σ_B^2).

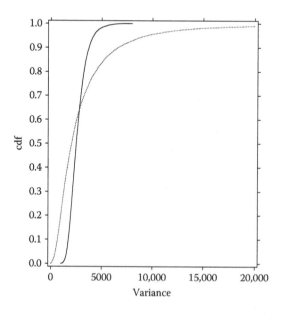

FIGURE 8.1
Variance components.

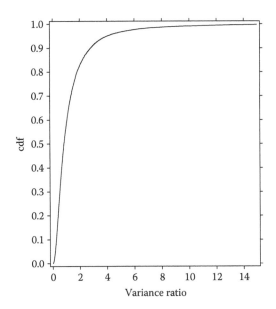

FIGURE 8.2
Variance component ratio.

The posterior median and mean of σ^2 are 2514 and 2660, and the 95% credible interval is [1493, 4676]. This agrees well with the exact interval [1495, 4744] from χ_{24}^2. (Some variation would be expected, since the draws with large σ^2 but small ϕ^2 giving a negative σ_B^2 have been deleted.)

The posterior median and mean of ϕ^2 are 2594 and 3811, and the 95% credible interval is [891, 13597]. The posterior is very diffuse, since it is based on only six batches. The posterior for σ_B^2 is correspondingly diffuse, with median 2072 and mean 3279, and 95% credible interval [330, 13008]. The posterior for the variance component ratio θ is shown in Figure 8.2; it is even more diffuse. The posterior median is 0.824, and the 95% credible interval is [0.102, 5.787]. The posterior for the overall mean μ is shown in Figure 8.3.

The posterior for the mean is also quite diffuse (relative to those for the individual batches), because it is also based on only six batches. The posterior median and mean are 1527, and the 95% credible interval is [1478, 1579].

Bayesian MCMC analyses of the dyestuff data are given in the WinBugs manual (1.4), using three different priors for the among-batch variance component. The posteriors for this variance component vary considerably across the priors, because of the small data information about this parameter. The posterior for σ^2 is not much affected. The priors are (1) uniform on σ_B^2, (2) uniform on the intraclass correlation $\sigma_B^2/(\sigma^2 + \sigma_B^2)$, and (3) gamma (0.001, 0.001) on the precision $1/\sigma_B^2$. The corresponding 95% credible intervals on σ^2 and σ_B^2, and their medians, are given in Table 8.2, with the flat priors on $\log \sigma$ and $\log \phi$ used above (4).

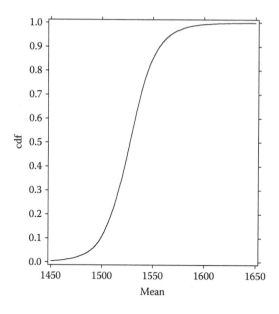

FIGURE 8.3
Yield mean.

For the individual batch means we show in Figures 8.4 through 8.9, four curves for each batch: the full Bayes posterior (solid curve), the EB posterior (dashed curve), the t posterior (dotted curve), and the overall mean (dot-dashed curve, as in Figure 8.3).

The figures show consistent patterns across batches. The t posteriors based on only the single batch mean (but using the pooled variance from all batches) are slightly less precise than the empirical or full Bayes posteriors. The latter are shrunk towards the overall mean, with the shrinkage and decreased precision for the full Bayes posteriors only slightly greater than for the empirical Bayes posteriors.

It is quite surprising that there is very little improvement by the empirical Bayes posteriors over the simple t-based posterior means, and it is also surprising that the EB posteriors, based on plug-in MLEs and the normal distribution, are only slightly different from the full Bayes posteriors, and

TABLE 8.2

Credible Intervals of 95% and Medians for σ^2 and σ_B^2

Prior	σ^2	σ_B^2
(1)	[1505, 4897], 2562	[264, 8658], 2306
(2)	[1515, 4680], 2538	[357, 9158], 1907
(3)	[1557, 5737], 2796	[0.009, 10290], 1306
(4)	[1493, 4676], 2514	[330, 13008], 2072

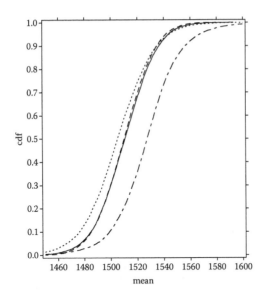

FIGURE 8.4
Batch 1 mean.

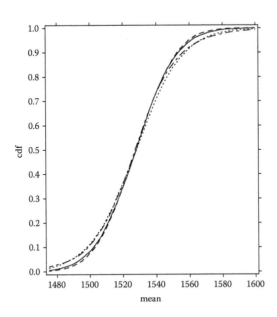

FIGURE 8.5
Batch 2 mean.

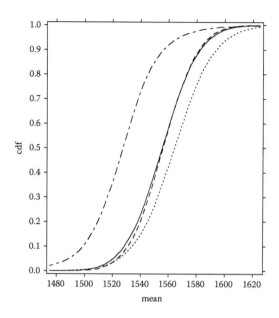

FIGURE 8.6
Batch 3 mean.

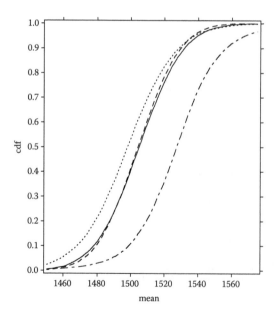

FIGURE 8.7
Batch 4 mean.

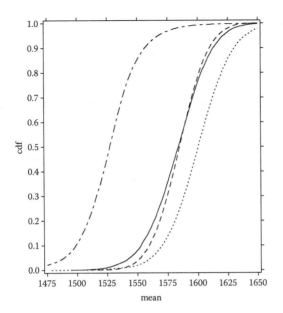

FIGURE 8.8
Batch 5 mean.

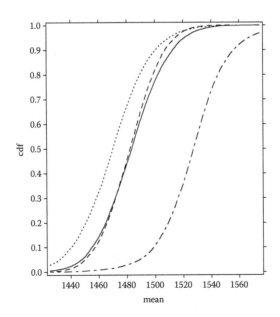

FIGURE 8.9
Batch 6 mean.

are only slightly over-precise. The small number of batches makes the batch variance component poorly defined, and this uncertainty is reflected in the considerable shrinkage and barely-increased precision of both empirical and full Bayes posteriors for the batch means. It reflects the price paid for increasing the model complexity with very limited information about the additional batch variance component parameter. The variance component model is at its most effective when there are *many* second-level units (batches here) with greater information within each batch.

The simple model analysis depends on the equality of batch sample sizes, and of their variances. The relaxation of these assumptions leads to more complex likelihoods, and (as in the frequentist framework) more complex analyses. For the case of unequal batch sample sizes, there are more sufficient statistics than parameters. For these more complex models, Bayesian analysis requires MCMC methods; we do not give details here.

8.3 Test for a zero variance component

An obvious question is whether we *need* the variance component model. Is there evidence of real variation in the batch means? The credible interval for σ_B^2 certainly suggests that it is nonzero; we now give the likelihood ratio for the null hypothesis $H_1 : \sigma_B^2 = 0$ against the alternative $H_2 : \sigma_B^2 > 0$. The likelihoods, likelihood ratio, and deviance difference are:

$$L_2(\mu, \sigma, \sigma_B) = c \cdot \frac{1}{\sigma^{r(n-1)}} \exp\left\{-\frac{W}{2\sigma^2}\right\} \cdot \left(\frac{1}{\phi}\right)^{r-1} \cdot \exp\left\{-\frac{B}{2\phi^2}\right\}$$

$$\cdot \frac{1}{\phi} \exp\left\{-\frac{r}{2\phi^2}(\bar{y} - \mu)^2\right\}$$

$$L_1(\mu, \sigma, 0) = c \cdot \frac{1}{\sigma^{r(n-1)}} \exp\left\{-\frac{W}{2\sigma^2}\right\} \cdot \left(\frac{\sqrt{n}}{\sigma}\right)^{r-1} \cdot \exp\left\{-\frac{nB}{2\sigma^2}\right\}$$

$$\cdot \frac{\sqrt{n}}{\sigma} \exp\left\{-\frac{nr}{2\sigma^2}(\bar{y} - \mu)^2\right\}$$

$$LR_{12} = \left(\frac{n\phi^2}{\sigma^2}\right)^{\frac{r-1}{2}} \cdot \exp\left\{-\frac{B}{2\phi^2}\left(\frac{n\phi^2}{\sigma^2} - 1\right)\right\}$$

$$\cdot \frac{\sqrt{n\phi}}{\sigma} \exp\left\{-\frac{r(\bar{y} - \mu)^2}{2\phi^2}\left(\frac{n\phi^2}{\sigma^2} - 1\right)\right\}$$

$$D_{12} = (r - 1)\log\left(\frac{\sigma^2}{n\phi^2}\right) + \frac{B}{\phi^2}\left(\frac{n\phi^2}{\sigma^2} - 1\right) + \frac{r(\bar{y} - \mu)^2}{\phi^2}\left(\frac{n\phi^2}{\sigma^2} - 1\right).$$

Simulation is straightforward, needing in addition to the random draws of σ^2 and σ_B^2, only independent random draws from $Z^2 = [r(\bar{y} - \mu)^2]/\phi^2 \sim \chi_1^2$.

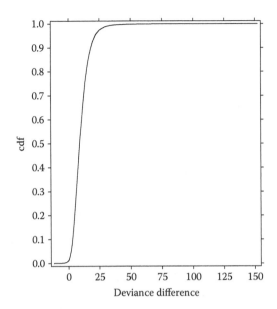

FIGURE 8.10
Deviance difference: variance component test.

Figure 8.10 shows the cdf from $M = 10{,}000$ draws from the posterior of D_{12} for the dyestuff example. Of the 10,000 draws, 135 are negative: the empirical posterior probability that the null model has a higher likelihood than the alternative model is 0.0135, with simulation standard error 0.0012. The 95% credible interval for the deviance difference is [0.76, 25.34], with median difference 8.95. The corresponding interval and median for the posterior probability of the null hypothesis, given equal prior probabilities, are [0, 0.406] and 0.0011. The posterior distribution of the posterior probability of the null hypothesis is very heavily skewed: though the median is 0.0011, the 97.5th percentile is quite high: 0.406. It is clear that the evidence supports the alternative, but the median does not represent the imprecision of the conclusion.

8.3.1 Alternative tests

There are both frequentist and Bayesian test alternatives to the posterior likelihood ratio. For the frequentist likelihood ratio test, the null hypothesis is on the boundary of the parameter space, so the usual asymptotic χ^2 distribution for the test statistic does not apply. The asymptotic distribution for this example is the "$\bar{\chi}$" distribution, a mixture of $1/2\chi_0^2 + 1/2\chi_1^2$, where χ_0^2 is the degenerate distribution with mass 1 at 0. The likelihood ratio test statistic is 5.42, which is compared for a 5% level test with $\chi_{1,0.10}^2 = 2.69$. The test firmly rejects the null hypothesis, with a p-value of 0.010.

The Bayes factor is difficult to use in this model because of the integration over the parameter space having to omit the region in which $\phi^2 < \sigma^2/n$. A solution to this difficulty was given by Pauler et al. (1999); for this example they give, with the same diffuse priors above, a log Bayes factor of -1.76, with simulation standard error 0.015, giving a posterior probability of the null hypothesis of 0.147, for equal prior probabilities.

The evidence against the null hypothesis for the Bayes factor is substantially weaker than from the likelihood ratio test, or the posterior likelihood ratio.

8.3.2 Generalized linear mixed models

These models, in which an exponential family response distribution is compounded with a random effect model in either overdispersion or variance component forms, generally require MCMC analyses. An exception is discussed in Aitkin et al. (2009), of a data set on lung cancer rates in Missouri cities, discussed originally by Tsutakawa (1985). Their discussion compares several different random effect models through their posterior deviance distributions, and also applies *Baysian model averaging* (Hoeting et al. 1999) over these random effect models to give robust posterior distributions for the city rates.

8.4   Finite mixtures

Finite mixtures have become a very important family of distributions since the ground-breaking paper on the EM algorithm (Dempster et al. 1977). We consider here the general K-component normal mixture model, with different means μ_k and variances σ_k^2 in each component:

$$f(y) = \sum_{k=1}^{K} \pi_k f(y|\mu_k, \sigma_k)$$

where

$$f(y|\mu_k, \sigma_k) = \frac{1}{\sqrt{2\pi}\sigma_k} \exp\left\{-\frac{1}{2\sigma_k^2}(y - \mu_k)^2\right\}$$

and the π_k are positive with $\sum_{k=1}^{K} \pi_k = 1$.

Given a sample $y_1, \ldots, y_n$ from $f(y)$, the likelihood is

$$L(\theta) = \prod_{i=1}^{n} f(y_i),$$

where $\theta = (\pi_1, \ldots, \pi_{K-1}, \mu_1, \ldots, \mu_K, \sigma_1, \ldots, \sigma_K)$. Maximum likelihood for the parameter vector θ is straightforward using an EM or other algorithm,

TABLE 8.3

Galaxy Recession Velocities (km/sec)

9.	17	35	48	56	78													
10.	23	41																
—																		
16.	08	17																
—																		
18.	42	55	60	93														
19.	05	07	33	34	34	44	47	53	54	55	66	85	86	86	91	92	97	99
20.	17	18	18	20	22	22	42	63	80	82	85	88	99					
21.	14	49	70	81	92	96												
22.	19	21	24	25	31	37	50	75	75	89	91							
23.	21	24	26	48	54	54	67	71	71									
24.	13	29	29	37	72	99												
25.	63																	
26.	96	99																
—																		
32.	07	79																
—																		
34.	28																	

provided that unequal standard deviations are bounded below. However, the global maximum of the likelihood may require detailed searching from multiple starting positions in complex models.

8.4.1 Example — the galaxy velocity study

We use a well-studied data set from Roeder (1990), adapted from more detailed data in Postman et al. (1986), to illustrate the posterior likelihood approach to an important mixture parameter: the number of components in the mixture. The data set and many Bayesian analyses of it were discussed in detail in Aitkin (2001), from which we adapt and extend the discussion below. The data in Table 8.3 are a stem-and-leaf plot of the recession velocities of 82 galaxies (in units of km/sec, given to 2 dp here) from six well-separated sections of the Corona Borealis region.

The question of interest is whether these velocities "clump" into groups or clusters, or whether the velocity density increases initially and then gradually tails off.

We investigate this question by fitting mixtures of normal distributions to the velocity data; *the number of mixture components* necessary to represent the data is the parameter of particular interest.

8.4.2 Data examination

Figure 8.11 shows the empirical cdf of the velocity data with a 95% simultaneous confidence band for the true cdf, and the superimposed single fitted normal cdf.

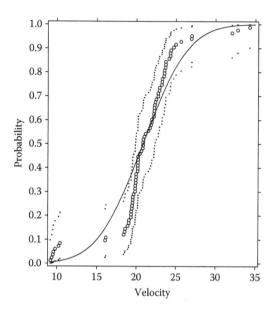

FIGURE 8.11
Normal distribution.

It is immediately clear that a single normal distribution is untenable, and the extreme tail observations on each side are clearly visible. However, the central part of the data looks reasonably normal and a three-component mixture may be sufficient.

8.4.3   Maximum likelihood estimates

Maximum likelihood estimates and frequentist deviances, for both equal and unequal variances, are shown for 1–7 components in Table 8.4. (The estimates for $K = 2$ given in Aitkin (2001) are for local maxima.)

For all models with $2 < K < 9$, the smallest 7 and largest 3 observations form stable components (note that $7/82 = 0.085, 3/82 = 0.037$). For $K = 4$ the central group of observations is split into two smaller groups with means around 22 and 20. Further increases in $K < 9$ fragment these two subgroups into smaller subgroups and shards. The evidence for three or four components looks quite strong. The fitted three-component mixture is shown in Figure 8.12 (solid curve, equal variances model; dashed curve, different variances model).

Aitkin (2001) and McLachlan and Peel (1997, pp. 194–196) used the frequentist bootstrap likelihood ratio test for the number of components. We do not give details here: as McLachlan and Peel noted, the test is nonconservative, with actual test size above the nominal level (because of the simulations from the fitted null model as though it was the true model), so the conclusions from the test are not well-supported.

TABLE 8.4

Parameter Estimates

K	k	μ_k Deviance	π_k	σ	μ_k Deviance	π_k	σ_k
1	1	20.83	1.000	4.54	20.83	1.000	4.54
		480.83			480.83		
2	1	21.88	0.953	3.03	21.43	0.788	3.04
	2	9.86	0.087		13.60	0.212	10.72
		461.00			413.78		
3	1	32.94	0.037	2.08	33.04	0.037	0.92
	2	21.40	0.877		21.40	0.878	2.20
	3	9.75	0.086		9.71	0.085	0.42
		425.36			406.96		
4	1	33.04	0.037	1.32	33.05	0.037	0.92
	2	23.50	0.352		21.94	0.665	2.27
	3	20.00	0.526		19.75	0.213	0.45
	4	9.71	0.085		9.71	0.085	0.42
		416.50			395.43		
5	1	33.04	0.037	1.07	33.05	0.036	0.92
	2	26.38	0.037		22.92	0.289	1.02
	3	23.04	0.366		21.85	0.245	3.05
	4	19.76	0.475		19.82	0.344	0.63
	5	9.71	0.085		9.71	0.085	0.42
		410.85			392.27		
6	1	33.04	0.037	0.81	33.04	0.037	0.92
	2	26.24	0.044		26.98	0.024	0.018
	3	23.05	0.357		22.93	0.424	1.20
	4	19.93	0.453		19.79	0.406	0.68
	5	16.14	0.025		16.13	0.024	0.043
	6	9.71	0.085		9.71	0.085	0.42
		394.58			365.15		
7	1	33.04	0.037	0.66	33.04	0.037	0.92
	2	26.60	0.033		26.98	0.024	0.018
	3	23.88	0.172		23.42	0.300	0.99
	4	22.31	0.221		22.13	0.085	0.25
	5	19.83	0.427		19.89	0.444	0.73
	6	16.13	0.024		16.13	0.024	0.043
	7	9.71	0.085		9.71	0.085	0.42
		388.87			363.04		

8.4.4 Bayesian analysis

The Bayesian analysis requires a full prior distribution specification for the mixture model parameters. These form a hierarchical set:

- The number of components K
- The proportions π_k in the K components
- The component-specific parameters θ, which are the μ_k and σ_k (or σ in the equal-variance model)

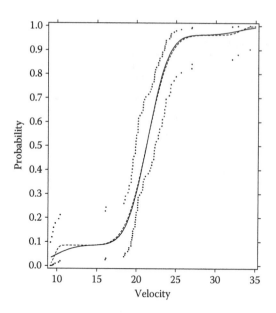

FIGURE 8.12
Three-component mixture.

Bayes analysis is, therefore, based on the full joint distribution of all the variables:

$$f(k, \pi, \theta, y) = f(k)f(\pi|k)f(\theta|\pi, k)f(y|\theta, \pi, k).$$

Analysis is greatly simplified by the introduction of a set of latent Bernoulli variables z_k for membership in component k; this allows the "complete data" representation

$$f^*(k, \pi, z, \theta, y) = f(k)f(\pi|k)f(z|\pi, k)f(\theta|z)f(y|\theta, z)$$

which allows simpler conditional distributions in the MCMC algorithm, as in the EM algorithm. For full generality an additional layer of prior structure can be added, in which the priors for k, π, and θ depend on independent hyperparameters λ, δ, and η, which then have to be specified.

(The analytic approach of Tan et al. (2010) through the inverse Bayes formula can be used here – it is given as Problem 3.7 on their pp. 114–115 – but the posterior requires a summation over all possible assignments of the observations to the mixture components, that is a sum over K^n terms. This is impractical for n of any size, though it could be reduced by eliminating assignments with negligible posterior probability.)

All the Bayes analyses used some form of data augmentation or Markov chain Monte Carlo analysis, with updating of the successive conditional distributions of each set of parameters (and the latent component membership variables) given the others and the data y. Most of the analyses took K initially

TABLE 8.5

Prior Distributions for K

K	1	2	3	4	5	6	7	8	9	10
EW	0.01	0.06	0.14	0.21	0.21	0.17	0.11	0.06	0.02	
CC	—	0.33	0.33	0.33	—	—	—	—	—	
PS	0.16	0.24	0.24	0.18	0.10	0.05	0.02	0.01		
RW	0.10	0.10	0.10	0.10	0.10	0.10	0.10	0.10	0.10	0.10
RG	0.03	0.03	0.03	0.03	0.03	0.03	0.03	0.03	0.03	0.030...
S	0.58	0.29	0.10	0.02	0.004	0.001	—	—	—	

as fixed and obtained an integrated likelihood over the other parameters for each K, and then used Bayes' theorem to obtain the posterior probabilities of each value of K.

More complex analyses (Richardson and Green 1997, Stephens 2000) used a form of MCMC in which K is included directly in the parameter space, which changes as K changes. Richardson and Green (1997) used reversible jumps (RJMCMC), and Stephens (2000) used a birth-and-death process (BDMCMC), to represent the transitions across different values of K.

The choice of prior distributions for K varied among the Bayes analysts. We report the priors for the analyses of the galaxy data by Escobar and West (EW1, EW2, 1995), Carlin and Chib (CC, 1995), Phillips and Smith (PS, 1996), Roeder and Wasserman (RW, 1997), Richardson and Green (RW, 1997) and Stephens (S, 2000). We also include the analysis by Nobile (N) in the discussion of Richardson and Green. These are summarized in Table 8.5.

The corresponding posterior distributions, using the hyper parameters specified by each set of analysts, are given in Table 8.6.

Posteriors vary widely across analysts, and most are very diffuse, with the striking exception of Roeder and Wasserman, whose posterior is almost a spike on $K = 3$ (with a prior uniform on 1–10).

8.4.5 Posterior likelihood analysis

For the posterior likelihood analysis, we report the analysis by Celeux et al. (2006). They used for each K a diffuse Dirichlet prior on the component

TABLE 8.6

Posterior Distributions for K

K	3	4	5	6	7	8	9	10	11	12	13
EW1	—	0.03	0.11	0.22	0.26	0.20	0.11	0.05	0.02	—	—
EW2	0.02	0.05	0.14	0.21	0.21	0.16	0.11	0.06	0.03	0.01	—
CC1	0.64	0.36	—	—	—	—	—	—	—	—	—
CC2	0.004	0.996	—	—	—	—	—	—	—	—	—
PS	—	—	—	0.03	0.39	0.32	0.22	0.04	—	—	—
RW	0.999	0.00	—	—	—	—	—	—	—	—	—
RG	0.06	0.13	0.18	0.20	0.16	0.11	0.07	0.04	0.02	0.01	0.01
S	0.55	0.34	0.09	0.01	—	—	—	—	—	—	—
N	0.02	0.13	0.16	0.25	0.20	0.13	0.06	0.03	0.01	0.01	—

proportions π_k, and diffuse conjugate priors on the means μ_k and inverse variances $1/\sigma_k^2$ (that is, the usual noninformative priors for the normal mean/variance models).

The MCMC sampler was run till convergence of the joint posterior distribution of the parameter set for each K. At convergence they sampled 10,000 values from the joint posterior distribution for each K, and computed the K-component mixture likelihood and deviance for each parameter set.

The object of their simulation was to evaluate various rules for *penalizing the posterior mean deviance* in the DIC of Spiegelhalter et al. (2002). The full deviance distributions were not used in their report. These were kindly provided by Chris Robert, and are shown for $K = 1, \ldots, 7$ initially in separate graphs, as the empirical and asymptotic (χ_{3K-1}^2 + frequentist deviance) distributions of the deviance (Figs. 8.13 through 8.19).

For $K = 1$, the asymptotic shifted χ_2^2 distribution (dotted curve) agrees almost perfectly with the empirical distribution (solid curve) – not surprisingly, since there *is* no mixture for $K = 1$. For $K = 2$ the shift agrees, but the shape of the empirical distribution is more diffuse than the asymptotic χ_5^2. For $K = 3$ the empirical distribution is far to the right of the asymptotic distribution, as well as being again more diffuse. This pattern is accentuated as K increases from 3 to 7.

This phenomenon has two aspects. First, as the number of components increases, the sample size per parameter decreases, so the posteriors for the individual parameters become more diffuse. This feeds into the posterior for the deviance, which therefore also becomes more diffuse – not just from the

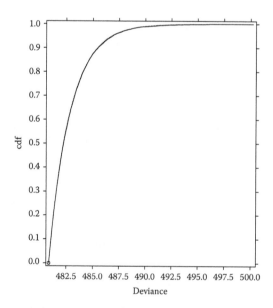

FIGURE 8.13
Deviance distributions for galaxy data, $K = 1$.

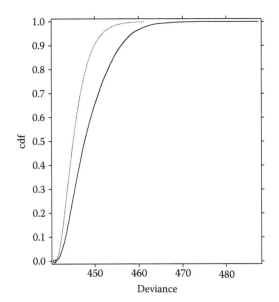

FIGURE 8.14
Deviance distributions for galaxy data, $K = 2$.

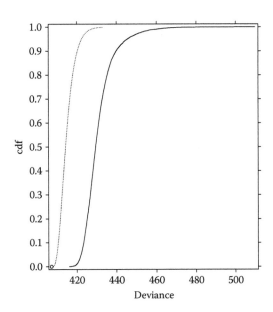

FIGURE 8.15
Deviance distributions for galaxy data, $K = 3$.

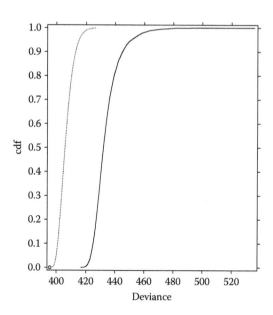

FIGURE 8.16
Deviance distributions for galaxy data, $K = 4$.

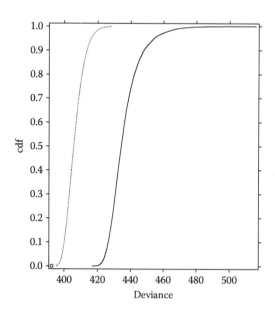

FIGURE 8.17
Deviance distributions for galaxy data, $K = 5$.

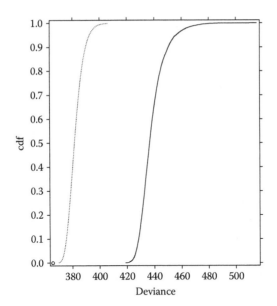

FIGURE 8.18
Deviance distributions for galaxy data, $K = 6$.

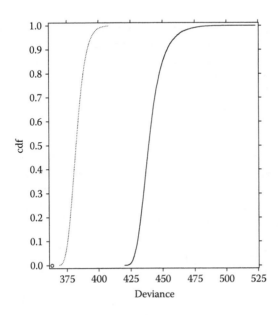

FIGURE 8.19
Deviance distributions for galaxy data, $K = 7$.

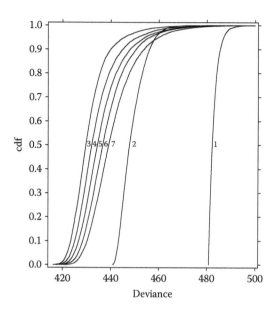

FIGURE 8.20
Deviance distributions for galaxy data, $K = 1$–7.

increasing degrees of freedom of χ^2, but also from the decreasing information for each parameter.

Second, as the dimension of the parameter space increases, the chance that a random parameter draw will be at or near the maximum likelihood estimate decreases, and the *achieved* simulation minimum of the deviance departs further from the analytic minimum.

These results emphasize the unreliability of conclusions based only on the MLEs and frequentist deviances for these models. This emphasis is even stronger when we place all the empirical deviance distributions on the same scale, in Figure 8.20.

The deviance distributions for $K = 1$ and 2 are far to the right of the others – the single normal and two-component mixture are very bad fits relative to the others. The distribution for $K = 3$ is the stochastically smallest – as K increases beyond 3 the deviance distributions move steadily to the right, to larger values (lower likelihoods). They also become more diffuse, with decreasing slope, as noted above.

Converting the deviance distributions to likelihoods and then posterior probabilities (for equal prior probabilities), we obtain the posterior probability distributions in Figure 8.21.

The median posterior probabilities for $K = 1, \ldots, 7$ are (to 3 dp) 0, 0, 0.297, 0.070, 0.018, 0.006, and 0.001. The evidence for $K > 2$ is very strong (note the median probabilities do not add to 1), but the 95% central credible intervals are very diffuse for larger K: [0, 0], [0, 0.002], [0, 0.995], [0, 0.982], [0, 0.954], [0, 0.867], [0, 0.684].

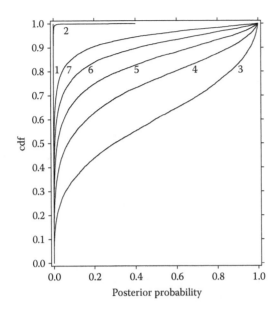

FIGURE 8.21
Posterior probability distributions for galaxy data, $K = 1$–7.

Thus the posterior probability distributions are not very helpful – a 95% credible interval for the probability of K components which stretches from 0 to almost 1 does not give useful information about this parameter. More compelling is the graph of the deviance distributions: adding additional components beyond three stochastically decreases the likelihood distribution, so we do not *need* more than three components to give a satisfactory representation of the galaxy velocities. This is the familiar issue of *parsimony* – there is no compelling evidence for more than three components, so we choose three as the parsimonious representation of the mixture. Celeux et al. (2006) came to the same conclusion: the variety of DIC penalties they examined *all* favored the three-component model.

We should note that Richardson and Green (1997) used graphs of the deviance distributions in one of their three examples (not the galaxy data). However, they gave kernel density estimates rather than cdfs, and it is much more difficult to identify stochastic ordering from estimated densities, especially if the bandwidth of the estimate is small, giving jagged estimates.

The mixture approach can be applied to other discrete parameter model comparison problems, for example, the number of neurons firing in motor neuron disease (see the discussion in Ridall et al. 2007).

8.4.6 Simulation studies

There are relatively few studies of competing Bayesian methods for assessing the number of mixture components. One by Steele and Raftery (2009)

TABLE 8.7

AIC and BIC for Galaxy Data

K	1	2	3	4	5	6	7
AIC	484.83	450.72	422.96	417.43	420.27	399.15	403.04
BIC	489.64	462.75	442.21	443.90	453.96	440.06	451.17

appeared in the last stages of preparation of this book. They compared six Bayesian methods on the galaxy data, and used data simulated from a standard normal density with five sample sizes, and from five two-component normal mixtures with varying sample sizes and mean and standard deviation differences. Two of the methods use penalties on the frequentist deviance: the Bayesian information criterion (BIC) uses the penalty $p \log n$ and the Akaike information criterion (AIC) uses the penalty $2p$, where p is the number of model parameters.

These are normally used as *decision* criteria – we choose the "best" model with the smallest value of BIC, or of AIC. These are simply calculated from the frequentist deviance, and are given in Table 8.7.

The "best" model by both AIC and BIC has $K = 6$. For AIC, $K = 7$ is the next best, while for BIC it is $K = 3$.

Steele and Raftery (2009) converted the BIC and AIC values to equivalent "integrated likelihoods" and computed posterior model probabilities (for uniform model priors) for each number of components from 1 to 7. AIC gives posterior probability close to 1 for $K = 6$, the remaining probability being on $K = 7$. BIC gives posterior probability 0.25 for $K = 3$ and probability 0.75 for $K = 6$.

The difficulty with these criteria is clear: they depend completely on the maximized likelihood and rely on the normality of the joint posterior distribution of the model parameters for all K around the MLEs. As is clear from the simulations described above, this assumption fails more seriously as K increases, but it is not valid even for $K = 2$ for the galaxy data. So no reliance can be placed on these criteria for samples in which the effective sample size per parameter is small.

Of the other four methods which used some form of integrated likelihood for the posterior probability calculation, one gave posterior probability 1 for $K = 3$, the others gave high probability for $K = 6$ and 7. Stephens (2000) used several truncated Poisson(λ) priors for K which gave considerably different results for $\lambda = 1, 3, 6$, and 25 (the prior in Table 8.5 is for $\lambda = 1$).

Steele and Raftery (2009) concluded that the BIC and Stevens's method performed best; since the simulation was limited to two-component mixtures and the sample sizes were not large, the relative performance of these methods and the posterior likelihood approach, with increasing sample sizes and increasing numbers of components, deserves further careful investigation.

References

d'Agostino, R. B., and M. A. Stephens. 1986. *Goodness of fit techniques*. New York: Marcel Dekker.

Agresti, A., and Y. Min. 2005. Frequentist performance of Bayesian confidence intervals for comparing proportions in 2 × 2 contingency tables. *Biometrics* 61:515–523.

Aitkin, A. G., and N. Hughes. 1970. *The parents' guide to coloured rod mathematics*. Sydney: Davies and Cannington.

Aitkin, A. G. 1975. *Number mastery*. Milton Queensland: Jacaranda Press.

Aitkin, A. G., and K. N. Green. 1984. *Maths skills*. Gladesville New South Wales: Shakespeare Head Press.

Aitkin, M. 1991. Posterior Bayes factors (with discussion). *Journal of the Royal Statistical Society B* 53:111–142.

Aitkin, M. 1997. The calibration of *P*-values, posterior Bayes factors and the AIC from the posterior distribution of the likelihood (with discussion). *Statistics and Computing* 7:253–272.

Aitkin, M. 2001. Likelihood and Bayesian analysis of mixtures. *Statistical Modelling* 1:287–304.

Aitkin, M. 2006. *Bayesian t-testing via the posterior likelihood ratio*. Unpublished.

Aitkin, M. 2007. *Bayesian goodness of fit assessment through posterior deviances*. Unpublished.

Aitkin, M. 2008. Applications of the Bayesian bootstrap in finite population inference. *Journal of Official Statistics* 24:21–51.

Aitkin, M., R. J. Boys, and T. Chadwick. 2005. Bayesian point null hypothesis testing via the posterior likelihood ratio. *Statistics and Computing* 15:217–230.

Aitkin, M. and R. Foxall. 2003. Statistical modelling of artificial neural networks using the multi-layer perceptron. *Statistics and Computing* 13:227–239.

Aitkin, M., B. J. Francis, and J. P. Hinde. 2005. *Statistical modelling in GLIM4*. Oxford: Clarendon Press.

Aitkin, M., B. J. Francis, J. P. Hinde, and R. E. Darnell. 2009. *Statistical modelling in R*. Oxford: Clarendon Press.

Aitkin, M., and C. C. Liu. 2007. *Bayesian hypothesis testing and model comparison via the posterior likelihood ratio*. Unpublished

Aitkin, M., C. C. Liu, and T. Chadwick. 2009. Bayesian model comparison and model averaging for small-area estimation. *Annals of Applied Statistics* 3:199–221.

Aitkin, M., and M. Stasinopoulos. 1989. Likelihood analysis of a binomial sample size problem. In *Contributions to probability and statistics: Essays in honor of Ingram Olkin*, ed. L. J. Gleser, M. D. Perlman, S. J. Press, and A. R. Sampson, 399–411. New York: Springer-Verlag.

Altham, P. M. E. 1969. Exact Bayesian analysis of a 2 × 2 contingency table, and Fisher's "exact" significance test. *Journal of the Royal Statistical Society B* 31:261–269.

Ankerst, D. P. 2005. Review of *Kendall's advanced theory of statistics Volume 2B: Bayesian inference* (2d ed.) by A. O'Hagan. *Journal of the American Statistical Association* 100:1465–1466.

Atkinson, A. 1985. *Plots, transformations and regression.* Oxford: University Press.

Banks, D. 1988. Histospline smoothing the Bayesian bootstrap. *Biometrika* 75:673–684.

Barnett, V. 1999. *Comparative statistical inference* (3d ed.). New York: Wiley.

Bartlett, M. S. 1957. A comment on D. V. Lindley's statistical paradox. *Biometrika* 44:533–534.

Bartlett, R. H., D. W. Roloff, R. G. Cornell, A. F. Andrews, P. W. Dillon, and J. B. Zwischenberger. 1985. Extracorporeal circulation in neonatal respiratory failure: A prospective randomized study. *Pediatrics* 76:479–487.

Begg, C. B. 1990. On inferences from Wei's biased coin design for clinical trials (with discussion). *Biometrika* 77:467–484.

Berger, J. O., and J. M. Bernardo. 1989. Estimating a product of means: Bayesian analysis with reference priors. *Journal of the American Statistical Association* 84:200–207.

Berger, J. O., B. Boukai, and Y. Wang. 1997. Unified frequentist and Bayesian testing of a precise hypothesis. *Statistical Science* 12:133–160.

Berger, J. O., B. Liseo, and R. L. Wolpert. 1999. Integrated likelihood methods for eliminating nuisance parameters. *Statistical Science* 14:1–28.

Berger, J. O., and L. R. Pericchi. 1996. The intrinsic Bayes factor for model selection and prediction. *Journal of the American Statistical Association* 91:109–122.

Bickel, P., and J. K. Ghosh. 1990. A decomposition for the likelihood ratio statistic and the Bartlett correction – a Bayesian argument. *Annals of Statistics* 18:1070–1090.

Binder, D. A. 1982. Non-parametric Bayesian models for samples from finite populations. *Journal of the Royal Statistical Society B* 44:388–393.

Box, G. E. P. and G. C. Tiao. 1973. *Bayesian inference in statistical analysis.* Reading, MA: Addison-Wesley.

Breiger, R. L. 1974. The duality of persons and groups. *Social Forces* 53:181–190.

Brewer, K. 2002. *Combined survey sampling inference.* London: Arnold.

Carlin, B. P., and S. Chib. 1995. Bayesian model choice via Markov Chain Monte Carlo methods. *Journal of the Royal Statistical Society B* 57:473–484.

Carota, C. 2008. Beyond objective priors for the Bayesian bootstrap analysis of survey data. Technical report.

Celeux, G., F. Forbes, C. P. Robert, and D. M. Titterington. 2006. Deviance information criteria for missing data models. *Bayesian Analysis* 1:651–674.

Congdon, P. 2005. Bayesian predictive model comparison via parallel sampling. *Computational Statistics and Data Analysis* 48:735–753.

Congdon, P. 2006a. Bayesian model choice based on Monte Carlo estimates of posterior model probabilities. *Computational Statistics and Data Analysis* 50:346–357.

Congdon, P. 2006b. Bayesian model comparison via parallel model output. *Journal of Statistical Computation and Simulation* 76:149–165.

Conigliani, C., J. I. Castro, and A. O'Hagan. 2000. Bayesian assessment of goodness of fit against nonparametric alternatives. *Canadian Journal of Statistics* 28:327–342.

Consonni, G., and L. La Rocca. 2008. Tests based on intrinsic priors for the equality of two correlated proportions. *Journal of the American Statistical Association* 103:1260–1269.

Cox, D. R. 1961. Tests of separate families of hypotheses. *Proceedings of the 4th Berkeley Symposium* 1:105–123.

Cox, D. R. 1962. Further results on tests of separate families of hypotheses. *Journal of the Royal Statistical Society B* 24:406–424.

Cox, D. R. 2006. *Principles of statistical inference.* Cambridge: University Press.

Cox, D. R. and Hinkley, D. V. 1974. *Theoretical statistics.* London: Chapman and Hall.

Cox, D. R., and N. Reid. 1987. Parameter orthogonality and approximate conditional inference (with Discussion). *Journal of the Royal Statistical Society B* 49:1–39.

Davis, A., B. Gardner, and M. R. Gardner. 1941. *Deep South*. Chicago: University of Chicago Press.

Dempster, A. P. 1974. The direct use of likelihood in significance testing. In *Proceedings of the conference on foundational questions in statistical inference*, eds. O. Barndorff-Nielsen, P. Blaesild, and G. Sihon, Hingham, MA: Kluwer, 335–352.

Dempster, A. P. 1997. The direct use of likelihood in significance testing. *Statistics and Computing* 7:247–252.

Edwards, A. W. F. 1972. *Likelihood*. Cambridge: University Press.

Ericson, W. A. 1969. Subjective Bayesian models in sampling finite populations (with discussion). *Journal of the Royal Statistical Society B* 31:195–233.

Escobar, M. D., and M. West. 1995. Bayesian density estimation and inference using mixtures. *Journal of the American Statistical Association* 90:577–588.

Ferguson, T. S. 1973. A Bayesian analysis of some nonparametric problems. *Annals of Statistics* 1:209–230.

Fisher, R. A. 1956. *Statistical methods and scientific inference*. Edinburgh: Oliver and Boyd.

Fox, J. -P. 2005. Multilevel IRT using dichotomous and polytomous response data. *British Journal of Mathematical and Statistical Psychology* 58:145–172.

Garthwaite, P., J. B. Kadane, and A. O'Hagan. 2005. Statistical methods for eliciting probability distributions. *Journal of the American Statistical Association* 100:680–701.

Geisser, S. 1984. On prior distributions for binary trials. *The American Statistician* 38:244–247.

Gelman, A., J. B. Carlin, H. Stern, and D. B. Rubin. 2004. *Bayesian data analysis* (2nd edn). Boca Raton: Chapman and Hall/CRC Press.

Gelman, A., X.-Li Meng, and H. Stern. 1996. Posterior predictive assessment of model fitness via realized discrepancies (with discussion). *Statistica Sinica* 6:733–807.

Ghosh, M. and G. Meeden. 1997. *Bayesian methods for finite population sampling*. London: Chapman and Hall.

Gönen, M., W. O. Johnson, Y. Lu, and P. H. Westfall. 2005. The Bayesian two-sample t test. *The American Statistician* 59:252–257.

Grice, J. V., and L. J. Bain. 1980. Inferences concerning the mean of the gamma distribution. *Journal of the American Statistical Association* 71:929–933.

Gutiérrez-Pena, E., and S. G. Walker. 2005. Statistical decision problems and Bayesian nonparametric methods. *International Statistical Review* 73:309–330.

Hartley, H. O., and J. N. S. Rao. 1968. A new estimation theory for sample surveys. *Biometrika* 55:547–557.

Henderson, H. V. and Velleman, P.F. 1981. Building multiple regression models interactively. *Biometrics*. 37: 391–411.

Herson, J. 1976. An investigation of relative efficiency of least squares prediction to conventional probability sampling plans. *Journal of the American Statistical Association* 71:700–703.

Hinde, J. P., and M. Aitkin. 1986. Canonical likelihoods: A new likelihood treatment of nuisance parameters. *Biometrika* 74:45–58.

Hjort, N. L., F. A. Dahl, and G. H. Steinbakk. 2006. Post-processing posterior predictive p-values. *Journal of the American Statistical Association* 101:1157–1174.

Hoadley, B. 1969. The compound multinomial distribution and Bayesian analysis of categorical data from finite populations. *Journal of the American Statistical Association* 64:216–229.

Hoaglin, D. C., F. F. Mosteller, and J. O. Tukey. 1985. *Exploring data tables, trends and shapes*. New York: Wiley.

Hodges, J. L., D. Krech, and R. S. Crutchfield. 1975. *StatLab: An empirical introduction to statistics*. New York: McGraw-Hill.

Hoeting, J. A., D. Madigan, A. Raftery, and C. T. Volinsky. 1999. Bayesian model averaging. *Statistical Science* 14:382–417.

Irony, T. Z., C. A. d. B. Periera, and R. C. Tiwari. 2000. Analysis of opinion swing: Comparison of two correlated proportions. *The American Statistican* 54:57–62.

Kahn, W. D. 1987. A cautionary note for Bayesian estimation of the binomial parameter *n*. *The American Statistician* 41:38–40.

Kass, R. E., and A. E. Raftery. 1995. Bayes factors. *Journal of the American Statistical Association* 90:773–795.

Kendall, M. G., and A. Stuart. 1996. *The advanced theory of statistics*. Vol. 3, 6. New York: Hafner.

Knorr-Held, L. 2000. Bayesian modelling of inseparable space-time variation in disease risk. *Statistics in Medicine* 19:2555–2567.

Kou, S. C., X. S. Xie, and J. S. Liu. 2005. Bayesian analysis of single-molecule experiments. *Applied Statistics* 54:469–506.

Lee, M. D. 2004. A Bayesian analysis of retention functions. *Journal of Mathematical Psychology* 48:310–321.

Lindley, D. V. 1957. A statistical paradox. *Biometrika* 44:187–192.

Lindsey, J. K. 1996. *Parametric statistical inference*. Oxford: Clarendon Press.

Lindsey, J. K. 1999. Some statistical heresies (with discussion). *Journal of the Royal Statistical Society D* 48:1–40.

Little, R. J., and D. B. Rubin. 1987. *Statistical analysis with missing data*. New York: Wiley.

Little, R. J. A., and H. Zheng (with discussion). 2007. The Bayesian approach to the analysis of finite population surveys. In *Bayesian statistics 8*, 1–20. Oxford: University Press.

Liu, C. C., and M. Aitkin. 2008. Bayes factors: Prior sensitivity and model generalizability. *Journal of Mathematical Psychology* 52:362–375.

Lohr, S. L. 1999. *Sampling: Design and analysis*. Pacific Grove: Duxbury Press.

McLachlan, G. J. 1987. On bootstrapping the likelihood ratio test statistic for the number of components in a normal mixture. *Applied Statistics* 36:318–324.

McLachlan, G. J. and Peel, D. 1997. *Finite mixture models*. New York: Wiley.

Newton, M., and A. Raftery. 1994. Approximate Bayesian inference using the weighted likelihood bootstrap. *Journal of the Royal Statistical Society B* 56:3–48.

Ng, K. W. 1997. Inversion of Bayes formulas: Explicit formulas for unconditional pdf. In *Advances in the theory and practice in statistics – A volume in honor of Samuel Kotz*, eds. N. L. Johnson and N. Balakrishnan. New York: Wiley.

Nicolae, D. L., X. -L. Meng, and A. Kong. 2008. Quantifying the fraction of missing information for hypothesis testing in statistical and genetic studies (with discussion). *Statistical Science* 23:287–331.

O'Hagan, A. 1995. Fractional Bayes factors for model comparisons (with discussion). *Journal of the Royal Statistical Society B* 57:99–138.

O'Hagan, A. 2004. Dicing with the unknown. *Significance* 1:131–132.

Olkin, I., Petkau, A. J., and Zidek, J. V. 1981. A comparison of *n*-estimators for the binomial distribution. *Journal of the American Statistical Association* 76: 637–642.

Owen, A. B. 1988. Empirical likelihood ratio confidence intervals for a single functional. *Biometrika* 75:237–249.

Owen, A. B. 2001. *Empirical likelihood*. Boca Raton: Chapman and Hall/CRC.

Pace, L., and A. Salvan. 1990. Best conditional tests for separate families of hypotheses. *Journal of the Royal Statistical Society B* 52:125–134.

Pauler, D. K., J. C. Wakefield, and R. E. Kass. 1999. Bayes factors and approximations for variance component models. *Journal of the American Statistical Association* 94:1242–1253.

Pearson, E. 1962. in the discussion of L. J. Savage – *The Foundations of Statistical Inference*, 55–56. London: Methuen.

Phillips, D. B., and A. F. M. Smith. 1996. Bayesian model comparison via jump diffusions. In *Markov chain Monte Carlo in practice* eds. W.R. Gilks, S. Richardson, and D. J. Spiegelhalter. London: Chapman and Hall.

Pierce, D. A. 1973. On some difficulties with a frequentist theory of inference. *Annals of Statistics* 1:241–50.

Pitman, E. J. G. 1965. Some remarks on statistical inference. In *Bernoulli, Bayes, Laplace*. eds. J. Neyman and L. M. LeCam, 209. New York: Springer-Verlag.

Pitman, E. J. G. 1979. *Some basic theory for statistical inference*. London: Chapman and Hall.

Postman, M., J. P. Huchra, and M. J. Geller. 1986. Probes of large-scale structures in the Corona Borealis region. *The Astronomical Journal* 92:1238–1247.

Richardson, S., and P. J. Green. 1997. On Bayesian analysis of mixtures with an unknown number of components (with discussion). *Journal of the Royal Statistical Society B* 59:731–792.

Ridall, P. G., A. N. Pettitt, N. Friel, R. Henderson, and P. McCombe. 2007. Motor unit number estimation using reversible jump Markov chain Monte Carlo (with discussion). *Journal of the Royal Statistical Society C* 56:235–269.

Robert, C. P., and George Casella. 1999. *Monte Carlo statistical methods*. New York: Springer-Verlag.

Roeder, K. 1990. Density estimation with confidence sets exemplified by superclusters and voids in the galaxies. *Journal of the American Statistical Association* 85:617–624.

Roeder, K., and L. Wasserman. 1997. Practical Bayesian density estimation using mixtures of normals. *Journal of the American Statistical Association* 92:894–902.

Royall, R. M. 1997. *Statistical evidence: A likelihood paradigm*. London: Chapman and Hall.

Royall, R. M., and W. G. Cumberland. 1981. An empirical study of the ratio estimator and estimators of its variance (with discussion). *Journal of the American Statistical Association* 76:66–88.

Rubin, D. B. 1981. The Bayesian bootstrap. *Annals of Statistics* 9:130–134.

Rubin, D. B. 1984. Bayesianly justifiable and relevant frequency calculations for the applied statistician. *Annals of Statistics* 12:1151–1172.

Rubin, D. B. 1987. *Multiple imputation for nonresponse in surveys*. New York: Wiley.

Ryan, T., B. Joiner, and B. Ryan. 1976. *Minitab students Handbook*. North Scituate, Mass: Duxbury Press.

Särndal, C. -E., B. Swensson, and J. Wretman. 1992. *Model-assisted survey sampling*. New York: Springer.

Savage, L. J. 1962. *Foundations of statistical inference*. London: Methuen.

Silvapulle, M. J. 1981. On the existence of maximum likelihood estimators for the binomial response models. *Journal of the Royal Statistical Society B* 43:310–313.

Spezzaferri, F., I. Verdinelli, and M. Zeppieri. 2007. Bayes factors for goodness of fit testing. *Journal of Statistical Planning and Inference* 137:43–56.

Spiegelhalter, D. J., N. G. Best, B. P. Carlin, and A. van der Linde. 2002. Bayesian measures of model complexity and fit (with discussion). *Journal of the Royal Statistical Society B* 64:583–639.

Squire, L. R. 1989. On the course of forgetting in very long-term memory. *Journal of Experimental Psychology: Learning, Memory and Cognition* 15:241–245.

StatXact. http://www.cytel.com/Products/StatXact/example_02.asp.

Steele, R. J., and A. E. Raftery. 2009. *Performance of Bayesian model selection criteria for Gaussian mixture models*. Technical Report 559, Department of Statistics, University of Washington.

Stephens, M. 2000. Bayesian analysis of mixtures with an unknown number of components — an alternative to reversible jump methods. *Annals of Statistics* 28:40–74.

Stone, M. 1997. Discussion of Aitkin (1997).

Tan, M. T., Guo-Liang Tian, and K. W. Ng. 2010. *Bayesian missing data problems: EM, data augmentation and noniterative computation*. Boca Raton: Chapman and Hall/CRC Press.

Tanner, M. A. 1996. *Tools for statistical inference* (3d ed.). New York: Springer-Verlag.

Tanner, M., and W. Wong. 1987. The calculation of posterior distributions by data augmentation. *Journal of the American Statistical Association* 82:528–550.

Tibshirani, R. 1989. Noninformative priors for one parameter of many. *Biometrika* 74:604–608.

Tsutakawa, R. K. 1985. Estimation of cancer mortality rates: A Bayesian analysis of small frequencies. *Biometrics* 41:69–79.

Valliant, R., A. H. Dorfman, and R. M. Royall. 2000. *Finite population sampling and inference: A prediction approach*. New York: Wiley.

Wald, A. 1947. *Sequential analysis*. New York: John Wiley & Sons.

Walker, S. G., and E. Gutiérrez-Pena. 2007. Bayesian parametric inference in a nonparametric framework. *Test* 16:188–197.

Ware, J. H. 1989. Investigating therapies of potentially great benefit: ECMO (with discussion). *Statistical Science* 4:298–340.

Welsh, A. H. 1996. *Aspects of statistical inference*. New York: Wiley.

Zellner, A. 1986. On assessing prior distributions and Bayesian regression analysis with g-prior distributions. In *Bayesian inference and decision techniques: Essays in honor of Bruno de Finetti*, ed. P. K. Goel and A. Zellner, 233–243. North-Holland: Elsevier.

Author Index

Subject Index